Australia's Wild Islands

The authors gratefully acknowledge the support
of Hanimex (Fuji Professional) and Vision Graphics.

Australia's Wild Islands

Foreword by Tim Flannery

QUENTIN CHESTER &
ALASDAIR McGREGOR

Hodder & Stoughton

Captions for photographs on previous pages:
Half-title page: Three Hummock Island.
Title page: The Hoskyn Islands in the Capricorn Bunker Group.
Opposite, top to bottom: Fruits of the Blue Quandong, Hinchinbrook Island; Granite shoreline, Three Hummock Island; Fitzroy Reef, Capricorn Bunker Group; Rock art of the Walmbaria people, Flinders Group.
Page vi: Waterfall on Hinchinbrook Island.
*Page viii: **T/C**.*

A Hodder & Stoughton Book

Published in Australia and New Zealand in 1997 by Hodder Headline Australia Pty Limited, (A member of the Hodder Headline Group) 10-16 South Street, Rydalmere NSW 2116

Copyright © 1997 by Quentin Chester and Alasdair McGregor

National Library of Australia Cataloguing-in-Publication data

Chester, Quentin.
Australia's wild islands.

ISBN 0 340 60537 5.

1. Islands - Australia. 2. Wilderness areas - Australia. I. McGregor, Alasdair, 1954- . II. Title.

919.40942

PHOTOGRAPHY
Unless otherwise indicated after the caption, all photographs were taken by Alasdair McGregor

PAINTINGS
All paintings are by Alasdair McGregor

MAPS
Lorenzo Lucia, Tech View Studio

DESIGN CONCEPT
Vivien Valk

DESIGN
Anna Soo

DEDICATION
To our fellow island travellers

Printed in Hong Kong

E N T S

FOREWORD

Islands are very special places for me. For most of my research career I have travelled to the islands of the south-west Pacific on fieldwork. I have discovered new species of possums, rats, bats and even tree-kangaroos in these places. While carrying out this work, I have been struck by how diverse, beautiful and fragile these islands often are. Australia's islands are similarly diverse, beautiful and fragile. Unfortunately, apart from a new notable exceptions, they are not as well known as the 'picture book' destinations of the south-west Pacific.

Evolution can take a distinctly different course on islands. They are often too small to support predators, so the creatures that inhabit them are fearless and to some extent helpless in the face of continental animals like ourselves. Their size also makes them vulnerable to change. A 'greenhouse' world will be one filled with drowned coral atolls. It may also be one which sees predominantly arid islands drenched with rain, and rainforest-covered islands parched by drought.

Despite their inherent fragility, the species inhabiting some islands have persisted there for millions of years. Such island endemics include the most bizarre, beautiful and ecologically informative creatures on earth.

Islands are often exceedingly beautiful places. You are never far from the sea, and mountains often rise abruptly from the coast. One rarely finds the monotony which can characterise large areas of the continents. Australia's islands are breathtakingly diverse, from the ice-capped volcano of sub-Antarctic Heard to the classic coral cays of the Great Barrier Reef. The selection presented in *Australia's Wild Islands* covers the full gamut of this diversity, brought to life in a way never achieved before.

We need to be reminded of the importance of our islands. Despite their beauty they are so often forgotten. When asked to name Australia's tallest mountain most Australians would suggest Mount Kosciuszko rather than Big Ben on Heard Island—about 250 metres taller.

Apart from their diversity, the wild islands which feature in this book have one thing in common: they have largely escaped the ravages that Europeans and their introduced fauna have committed on continental Australia. This makes them precious beyond reckoning. Their story is beautifully told here. Quentin Chester and Alasdair McGregor's text contains a most sensitive combination of historical reference and personal encounter. At the same time their text is informative, including a sizable body of scientific information, presented in an easily understood manner.

Visually, this book does great justice to its subject. Photographically it adds up to one of the most vibrant celebrations of Australian nature published in recent years. It is also graced with some extraordinary artwork. Alasdair McGregor is one of Australia's finest natural history artists and his paintings in this book are high points, adding an extra insightful dimension to the visual feast. They tell the story of island life in a way that words and photographs alone never could.

Most Australians will never have the opportunity to visit even a small fraction of the islands featured here, for most are difficult to reach and some have access restricted in order to protect wildlife. This magnificent book does a great service in bringing the beauty, diversity and value of those remote islands within reach of us all.

DR TIMOTHY FLANNERY
PRINCIPAL RESEARCH SCIENTIST
AUSTRALIAN MUSEUM, SYDNEY

INTRODUCTION

The Australian continent is the largest member of an archipelago that is some 12,000 strong. Including those flanking its Antarctic Territory, Australia's islands are dispersed across more than 11 million square kilometres of ocean, spanning 60° of latitude and 72° of longitude. Within this Australian 'archipelago' there is an astonishing diversity of offshore habitats—arguably the greatest of any country in the world.

For many Australians, the mention of islands brings to mind those plots of paradise enclosed by the Great Barrier Reef or perhaps the rocky sentinels flanking Bass Strait. Some might think of the larger examples, like Kangaroo, Bathurst and Melville. But the vast majority of the continent's outliers languish in obscurity. The fictional locales visited by all manner of castaways from Gulliver to Gilligan are better known than many of the isles dotted around our own shores.

How many Australians are aware that their tallest mountain, only glaciers and sole active volcano are found on one of their sovereign islands? Or that the sites of the country's biggest resource development, largest body of fresh water and richest iron ore deposits all lie offshore? Or that islands are responsible for Australia's first export industries, its earliest shipwrecks and largest bird populations?

For a nation whose populace dwells overwhelmingly on the coastal 'verandah' this neglect is surprising. Australians are often accused of having their backs turned to the forbidding interior of the continent. In another sense, it seems we rarely lift our gaze from the sand and surf to consider what lies beyond. Like the inhabitants of ancient Sybaris, Australians are a people captivated by the delights of their near shores.

Tropical to Temperate

This book aims to present some striking aspects in the life and times of a small sample of Australian islands. We have tried to give an outline of each island's natural and cultural history. At the same time there is an attempt to convey a sense of 'being there' through our personal responses to place and experience.

The main criterion in selecting the 24 islands and archipelagos featured here was to reflect the remarkable diversity of Australia's offshore domains. Of the continental examples, 10 are located in the tropics and the remainder distributed across sub-tropical and temperate waters.

The contrasts are bewildering. There are mountainous hulks of islands over 2,500 metres tall and wafers of land that barely clear high tide. Some locations receive an annual rainfall exceeding 2,000 millimetres; others are lucky to get a tenth of that figure. The nearest is a mere toss of a stone from the mainland, while the most remote lies more than 4,000 kilometres distant. Many are cast-off continental relics, marooned by the rising seas at the end of the last ice age. Others are remnant volcanoes or ever-changing accretions of coral and sand.

Crucibles of Evolution

But the significance of these islands goes far beyond being a 'chocolate box' of curiosities and marvels. At the time when many of the places featured in this book were being prevailed upon by sealers and settlers, a young Charles Darwin was diligently studying the denizens of the Galápagos Islands off the coast of Ecuador. He noted the odd behaviour of the world's only seafaring lizard, and variations among creatures as diverse as finches and giant tortoises. Meanwhile, half a world away, Alfred Russell Wallace devoted eight years to making similar observations in the Malay archipelago. From their independent island research these two men laid the foundation of evolutionary theory and biogeography.

Had Darwin or Wallace been waylaid on

Australian islands they would have found equally compelling evidence for their radical theories. In the often simplified world of island environments the processes of evolution are accentuated. Speciation and extinction can happen over relatively short periods of time. These dynamics make islands living laboratories.

The value of such study is more than theoretical. Australia's islands are refuges for myriad endemic species. Many islands are also life-rafts for plants and animals that are endangered and extinct on the mainland. For example, Australia's parlous record of mammal extinctions would be much worse were it not for the existence of these exiled populations. Some nine mammal species gone from the mainland are still extant on islands. The survival of a further 16 endangered or vulnerable mammals may well depend on the existence of their offshore relatives.

The transformation of Australia's landscapes that has contributed to this dire history is now at a stage where native flora and fauna is increasingly consigned to relatively small parcels of habitat. Many of our parks, reserves and so-called wilderness areas are themselves virtual islands, surrounded by expanses of agricultural and urban development.

Within such mainland fragments, the survival of species becomes precarious due to a critical loss of range and habitat, and a dwindling genetic diversity. They are insulated from the benefits of dispersal and immigration that help keep an ecosystem in equilibrium. At the same time these patches are vulnerable to the threat of introduced species, diseases, and natural and man-made disasters. The lessons gleaned from offshore islands are fundamental to the conservation of the continent at large.

Exiles and Exports

While our outliers can be viewed as natural repositories, they also played a crucial part in Australia's human history. They were stepping stones that helped bring people to these shores for the first time. For Aboriginal Australians, islands have always been a source of bountiful harvest and spiritual power. And intrepid coastal voyages made by Aborigines in small canoes, rafts and catamarans rank with the greatest maritime feats of any age.

For the first Europeans to visit the continent, these islands were poles of reference in the quest to resolve the mysteries of Terra Australis. It is impossible to sail our coastline and not be in awe of navigators like Tasman, Cook, Flinders, Baudin and King. In their hazardous journeys islands were often places of sustenance and refuge. At other times they were seen as desolate and forbidding. For the hapless victims of shipwrecks they could also be sites of terrible ordeals.

In the early days of white settlement, the sealing and whaling industries made islands central to the economy. During these decades avarice, hope and misery often found their keenest expression offshore. The human suffering experienced at infamous convict outposts like Norfolk, Sarah and Pinchgut was rivalled only by that endured by dispossessed Aboriginal groups, on islands such as Bernier and Dorre, Fraser, and Flinders in Bass Strait.

Islands continue to exert a powerful, but often unsung, influence on our lives. One need think only of the abalone and rock lobsters hauled from our island environs, or the iron ore, manganese and oil extracted offshore. And where would Australian tourism be without such places, both in their unadorned state and as confected dystopias of palms and pools? And what of the ramifications caused by one Torres Strait Islander initiating a legal challenge to the notion of 'Terra Nullius'?

Many Isles, One World

No matter how remote these islands may seem, none has remained free from human intrusion. Given this legacy of disturbance, defining what makes an island 'wild' remains highly subjective.

There are also many alternatives equally deserving of inclusion as those featured here. Islands like Barrow, Raine, Lizard, Cape Barren, Maria, Pearson are, by any measure, outstanding. These, and any number of other examples, could have easily been included without straying from our criterion. But then this book may well never have been finished—certainly not in this century.

Part of the universal appeal of islands is their isolation. To be a castaway on a desert island remains one of the most compelling fantasies in western culture. Such places satisfy a deep yearning for seclusion and freedom. To be able to walk around an island, to be a party to all its facets and moods, is like meeting one of our own. Whereas continents remain essentially unknowable, an island is often a landscape on a human scale.

At the same time, however, it is essential to recognise that these places are wedded to their surroundings, both near and far. When John Donne wrote that "no man is an island", such locales were the epitome of separateness. But the world and how it is perceived has changed a lot in the past 400 years. Connected by wind and waves, they are inextricably bound up with changes mankind has wrought on the planet.

Island Life

Walking these lonely shores was often exhilarating. Invariably, however, there were jarring reminders of what lay beyond the horizon. Every island, no matter how far flung, had its share of flotsam. As well as turning up shells, driftwood and quaint fishing floats, the modern beachcomber has to be prepared for a sorry mélange of plastic bottles and bags, nylon rope, polystyrene, cans and packing straps.

The winds responsible for delivering this debris were a persistent presence in our days of self-imposed exile. Islands are blustery, buffeted places. We were often grateful for the moderating influence of sea breezes. At other times we copped a fearful battering.

But even when the elements were at their most unruly we were rarely alone. Nothing quite prepared us for the sheer profusion and exuberant presence of so much wildlife. It always came as a jolt to be suddenly in the midst of a multitude of squawking, stinking seals and penguins, or to come under the inquisitive gaze of wallabies or bettongs. But for all the varied creatures we witnessed there is no doubt that, ultimately, islands belong to birds. In their 'oneness' with place and their mastery of the elements, these winged hosts were an inspiration: palpable symbols of the links between air, land and sea.

For the most part the birds have these islands to themselves. We did, nevertheless, visit a few outposts where humans were also in residence. It was a privilege to meet a few of those who have chosen to live offshore. These island dwellers exhibited a resourcefulness and sardonic humour, mixed with a healthy disregard for society, that was oddly reassuring.

Typically, these islanders were protective but not possessive of their patch. That such places should elicit these responses is perhaps further evidence of islands being seen as both approachable and at the same time vulnerable. With a continent at large, the dimensions of the land and its manifest ills can be overwhelming; on islands the boundaries, problems and opportunities are often more readily defined.

If Australia itself is to continue to be one of the world's great 'wild islands,' then perhaps islands can help us attend to what we still have and what is at risk. So much of the history associated with the sites discussed in this book involves an almost unbearable sense of loss. Yet islands continue to be places of hope and often surprising resilience. Our accounts are mostly devoted to the challenge of looking in on what appear to be self-contained worlds, but the view looking out should be equally compelling. No man is an island, and nor, in the end, are islands truly solitary. They are singular realms yielding multiple perspectives. With that vision comes the possibility of a new kind of belonging.

QUENTIN CHESTER
AND ALASDAIR McGREGOR

LOCATION OF ISLANDS
AND ISLAND GROUPS

ARAFURA SEA

Melville Island

Bathurst Island

Torres Str

Wessel Islands

Bus

*GULF
OF
CARPENTARIA*

Darwin

Bonaparte
Archipelago

Sir Edward Pellew Group

CAPE
YORK

Cookte

Wyndham

Broome

INDIAN OCEAN

Dampier Archipelago

Port Headland

Bernier and Dorre Islands

Carnarvon

Houtman Abrolhos

Geraldton

Ceduna

Nuyts Archipelago

Fremantle

Esperance

Adelaide

Archipelago of the Recherche

GREAT AUSTRALIAN BIGHT

Kangaroo Island

SOUTHERN OCEAN

Melbour

Glennie Group

King Island

BA
STR

Hunter Group

Hoba

Maatsuyker Group

GREAT

ers Group

BARRIER

CORAL SEA

Hinchinbrook Island

asville

REEF

PACIFIC OCEAN

Whitsunday Islands

Capricorn Bunker Group

Gladstone

Fraser Island

Brisbane

Port Macquarie

Lord Howe Island

Cabbage Tree Island

Sydney

Spectacle Island

TASMAN SEA

Kent Group

Flinders Island

Christmas Island

Australia

Macquarie Island

Heard Island

Antarctica

South America

Africa

Towards Half Moon Reef (Pelsaert Island, Houtman Abrolhos)
Oil on canvas 107 x 91cm

ARKS AND EXILES

HOUTMAN ABROLHOS
BERNIER AND DORRE ISLANDS
DAMPIER ARCHIPELAGO

*Paradise has many guises. To the untutored
eye these islands can appear arid and
cheerless. Yet to seabirds they offer secure
nesting sites. For a host of endangered small
mammals there is immunity from feral
predators. Seen on their own terms, these are
among Australia's most treasured islands.*

HOUTMAN ABROLHOS

Abri Vossos Olhos

Looking past the limpid shallows enclosed by Half Moon Reef, a distant ribbon of surf glinted in the afternoon sun. It would have been easy to forget that I was standing on the site of one of the most remarkable episodes in Australia's maritime history. Except for the rowdy south-westerly breeze, the apparent sanctuary of that broad lagoon belied the fear that has become synonymous with the scattering of islands, reefs and shoals known as the Houtman Abrolhos. For this is a dangerous place; its very name sounds alarm.

The word "Abrolhos" is a corruption of the Portuguese *abri vossos olhos*, a warning cry of mariners for centuries. It came to signal "spiked or rocky obstructions" but in translation is literally, "keep your eyes open".[1]

I was standing on Gun Island, typical of most in the scattering of more than 100 that make up the Houtman Abrolhos: low, flat and treeless, rising no more than a few metres out of the lagoon. Many of the islands in the archipelago are no bigger than suburban building blocks. These coral and limestone islets lie atop part of a drowned coastline at the edge of the continental shelf, 60 kilometres west of Geraldton. The underlying platforms on which they ride consist of about 900 metres of Cretaceous and Tertiary limestone and siltstone, first laid down more than 100 million years ago. In the most recent sequence of sea level change, separation from the mainland occurred only within the last 11,000 years.

Together with solitary North Island, the Pelsaert, Easter and Wallabi Groups — three clusters of islands that make up the Houtman Abrolhos stretch for 80 treacherous kilometres, their margins exposed to the full expanse of the Indian Ocean.

Over the past 10,000 years, coral reefs have developed on the northern and eastern sides of the groups, attaining their present configuration some 4,000 years ago. The presence of such an extensive reef system at high latitudes is due entirely to the Leeuwin Current, a warm offshore flow that drifts south along the coast of Western Australia past the Houtman Abrolhos, finally petering out near Cape Leeuwin. Water temperatures are on average 2.5 to 2.7° Celsius higher than would otherwise be expected for such latitudes. Bathed in warm water through winter, corals flourish near the limit of their development. The Houtman Abrolhos supports the most southerly aggregation of coral reefs in the Indian Ocean. Its reefs are also among the most diverse in the world. For centuries these underwater gardens bloomed in deadly readiness.

In the early evening of 9 June 1727, the Dutch East India Company (VOC) ship *Zeewijk* struck Half Moon Reef, approximately six kilometres west of Gun Island.[2] The *Zeewijk* was en route via the Cape of Good Hope to Batavia, centre of Dutch power on Java. Nearly three weeks earlier in mid-ocean, the *Zeewijk*'s captain, Jan Steyns, had inexplicably decided to change course and steer east-north-east towards the coast of Western Australia, known to the Dutch at that time as Eendracht's Land.[3] Steyns' decision, perhaps an act of foolhardy curiosity, proved disastrous.

The ship was pounded by surf for more than

a week before a boat could be launched. Of the 158 men that had left the Cape, 96 managed to struggle to the relative safety of the 800 metre-long Gun Island.[4] Lucky to survive the trials of the reef, the castaways must have thanked providence when they discovered freshwater. They had by chance been marooned during winter, when storms bring most of the region's annual rains.[5] Riven with sink holes, the island and others nearby brimmed with potable water.

A small party was quickly dispatched in the ship's surviving longboat for Batavia. Several months passed. Unbeknown to the remaining castaways, the *Zeewijk*'s longboat was never sighted again after it left Gun Island. They assumed the worst and set to work, building a yacht from timbers salvaged from the *Zeewijk*. Hardwood ribs and knees were cut from mangroves found across the lagoon on Pelsaert Island, and nearly five months after laying its keel, 88 men departed the Houtman Abrolhos in the little sloop they affectionately named *Sloepie*. The *Sloepie* proved to be sturdy and seaworthy. At the end of April 1728, ten months after striking Half Moon Reef, 82 survivors reached Batavia, carrying an intact fortune in coins from the *Zeewijk*.

"... fire and blood and lace and velvet and steel ..."[6]

The dangers of the Houtman Abrolhos had been known for over 100 years before the *Zeewijk* foundered. Seventeenth-century navigation had no accurate means of calculating longitude, so it was with some trepidation that Dutch ships plying the fast route to the East Indies approached the uncharted coast of Western Australia.

A fleet commanded by Frederick de Houtman, after whom the islands are known, came upon the group in 1619: "... at night ... we again came upon a low-lying coast, a level country with reefs all around it. We saw no highland or mainland, so that this shoal is to be carefully avoided as very dangerous to ships that wish to touch at this coast ..."[7]

A second fleet almost met with disaster in 1627. Then, in the early hours of 4 June 1629, the inevitable happened.[8] The *Batavia*, flagship of a grand fleet of six VOC ships, ran aground on Morning Reef in the Wallabi Group, precipitating what remains as one of the most infamous episodes in the history of Europeans in Australia. The *Batavia*'s maiden voyage ended abruptly when 316 men, women and children were cast onto the reef.

After several days, most of the 276 survivors had struggled ashore on two small islands close to the wreck. But rather than enforce discipline, the fleet commander, Francisco Pelsaert, departed in the ship's boat to look for water on the mainland coast. The hunt proved fruitless and some 500 kilometres north of his marooned companions, Pelsaert decided to press on for help rather than return. Accompanied by the *Batavia*'s master and 46 passengers and crew, he sailed directly for Java.

Before the shipwreck, mutiny had been festering on the *Batavia* as she made passage from the Cape of Good Hope. Secret plans had been well advanced for the seizure of the ship, and its rich cargo, the women and wine. Commander Pelsaert's absence was the opportunity the mutineers craved. A drunken rabble embarked on a frenzy of murder, rape and debauchery, led by a charismatic psychopath, VOC merchant officer, Jeronimus Cornelisz.

Pelsaert's passage to Batavia in the open ship's boat took only 33 days but his return in the yacht *Sardam* was slow and tedious. More than three months had passed. Pelsaert's return coincided with an attack by the mutineers on a poorly armed garrison of loyal soldiers and survivors ensconced on West Wallabi Island. The *Sardam* itself was very nearly seized, but Wiebbe Hayes, leader of the garrison, frantically rowed out to the ship and alerted Pelsaert in the nick of time. The miscreants were quickly subdued.

Fearing further insurrection, Pelsaert hastily convened formal trials. Cornelisz and his accomplices were tortured according to Dutch law, and confessions extracted. Seven were then mutilated and hanged *in situ*. Two of the guilty were released on the mainland to meet their fate with the Aborigines or the elements. Another five were executed when the *Sardam* reached Batavia. In between dispensing justice, Pelsaert was able to direct the salvage and

recovery of much of the treasure languishing on the wreck of the *Batavia*. Eight of the ten chests of bullion were retrieved.[9]

Of those aboard the *Batavia* when it ran onto Morning Reef, only 116 remained alive when the *Sardam* reached Java in December 1629. Six months of total perdition off the alien coast of Eendracht's Land were at last over. Pelsaert died within two years, destroyed by guilt. Two hundred killed, a fine new ship wrecked and much cargo lost — for the VOC the *Batavia*'s first and only voyage was a political and economic disaster.

The ghosts of other Dutch wrecks may well haunt the Houtman Abrolhos. John Lort Stokes and John Wickham in HMS *Beagle* made the first detailed survey of the group in 1840, and postulated the location of the *Batavia*. Sighting wreckage at the extreme corner of the most southerly group, Wickham named the island after *Batavia*'s commander — erroneously as later searches and detailed study of Pelsaert's journals revealed. The *Batavia* was discovered in 1963 at the southern end of the Wallabi Group.[10] The identity of Wickham's wreck remains a mystery; no trace has yet been found.

Roosting Common Noddies squabble atop a thicket of Nitraria billardierei *on Pelsaert Island, the only breeding locale for this species in the Houtman Abrolhos. More than 130,000 pairs nest on open ground at the southern end of the island each spring and summer. After breeding they put to sea in vast flocks.*

Seabird Islands

From Wreck Point, near the site of the "phantom" *Batavia*, it is an easy slosh across the reef platform at low tide to tiny Jon Jim Island. Dismembered parts — boilers and encrusted ironmongery — are cast about the reef. *Jon Jim*, a wooden freezer boat, ran onto the reef in 1961. It was the twentieth vessel known to have foundered on or near the Houtman Abrolhos in the 375 years since Houtman's sighting.

Roughly circular and barely 50 metres across, Jon Jim Island is, I found, home to a colony of more than 100 breeding pairs of Roseate Terns (*Sterna dougallii*). Nesting behind a wall or eroded limestone, the colony lifted almost as one, beating into a stiff Abrolhos breeze. Brilliant-white undersides shone, flushed with the faintest pink on the breast, their flowing tail streamers and vermilion feet dazzling against a cerulean sky. As quickly as they rose they dropped, squabbling for a while before settling over their eggs. Their nests are nothing more than a scrape in the coral rubble lined with the odd scrap of local vegetation.

The Roseate Tern's world-wide distribution is broad, although in Australian waters it is quite rare, breeding only off the west coast and to a limited extent on Great Barrier Reef cays. The Houtman Abrolhos is its stronghold in Western Australia, with more than 3,400 pairs throughout the archipelago.

Just as tiny Jon Jim Island provides precious breeding space, so do the rest of the islands, for above all else they are a place of birds — seabirds in their millions, all profiting from the riches of the surrounding waters.

Some 19 species, over 1,600,000 seabird pairs nest on 122 islands, islets and rocks.[11] The most numerous — the Wedge-tailed Shearwater (*Puffinus pacificus*) — excavates nest burrows where sand has accumulated in permanent dunes. Over one million pairs nest on West Wallabi alone, with a further 75,000 in the sand beds and coral grit of the southern end of Pelsaert Island.

Past the shallows of the reef terrace, broadcast with concretions of dead coral, Pelsaert Lagoon unfolds in a swirl of blues, bands of aquamarine diving to the deepest cobalt. A line of vivid white cuts the horizon in the middle distance, where a narrow beach on Pelsaert Island skirts the lagoon. Flat and unrelieved, Pelsaert appears to barely struggle out of the sea.

I stayed on Jon Jim Island watching the Roseate Terns for nearly two hours. In that time the tide had turned and now began to wash around its undercut rim. It was time to go. As I crossed back to the "mainland", clouds of birds rose from their distant roosts among rambling metre-high thickets of the succulent, sage green shrub *Nitraria billardierei*, common on the dunes at the southern end of Pelsaert Island. Hundreds took to the air, spreading like smoke over the lagoon. Dozens came my way, croaking and swerving just overhead, their shadows dancing over beach and shallows.

Most were the curiously named Common Noddy (*Anous stolidus*), a tern of tropical and subtropical waters.[12] The Common Noddy used to nest in stupendous numbers on Rat Island in the Easter Group. Early accounts estimated up to 700,000 pairs on an island only one kilometre long. John Lort Stokes remarked on its rich guano deposits, a legacy of the noddies and equally astounding numbers of Wedge-tailed Shearwaters and Sooty Terns (*Sterna fuscata*). Today, the total bird population of all species comprises just two Osprey (*Pandion haliaetus*) pairs — not a single noddy, shearwater or tern remains.

Black Rats (*Rattus rattus*), castaways from some unknown early shipwreck, led Stokes to name the place Rat Island. The rats, however, were not singularly responsible for the decimation of the birdlife. In 1840 Stokes reported on "soil mixed with guano and filled with burrows of the sooty petrel",[13] and by 1847 guano miners had been attracted to Rat, Pelsaert, Gun and several smaller islands.

The miners stripped away vegetation and topsoil. Nothing but bedrock remained; no place for the burrowing shearwater. The guano diggers dined on birds' eggs, and the cats they brought to Rat Island for rodent control preyed on the birds as well. Mining lasted on Rat Island until 1915 and, with the cats and rats in deadly combination, all the birds were gone by the 1940s. Although Pelsaert was also mined, its seabird populations largely escaped the catastrophe that was Rat Island.[14]

*Above: A Crested Tern (*Sterna bergii*) returns to its chick after a successful fishing trip. Noisy colonies of a few pairs to gatherings of more than a thousand nests are found on sandy ground across nine Abrolhos islands.*

Opposite: Lesser Noddies nesting among White Mangroves on Pelsaert Island. A fluctuating population approaching 50,000 pairs nests in the Houtman Abrolhos, their only breeding station in Australian waters. Nesting in spring and summer, Lesser Noddies generally remain close to their colonies year round.

*Pied Cormorants
(*Phalacrocorax varius*) *nest
in a large colony of about
1,000 pairs on Wooded Island
in the Easter Group.*

I walked north via the lagoon beach, drifts of tiny white shells crunching underfoot. Noddies and Sooty Terns barked and croaked overhead. Passing the forlorn yet rather graceful sag of Pelsaert Island's ruined guano-loading jetty, a low wind-pruned stand of mangroves came into view less than a kilometre further north.

Tightly drawn on the shore and rising barely more than a couple of metres above the coral rubble piled behind, a forest of White Mangroves (*Avicennia marina*) twitched with life. Birds darted out in all directions from its canopy. Others perched defiantly on top branches, the thrashing breeze unable to shake them free. They were Lesser Noddies (*Anous tenuirostris melanops*), an endemic subspecies restricted to the Houtman Abrolhos.[15]

I pushed my way into the depths of this diminutive forest, my feet slowly sinking into malodorous mud. Tiers of nests decorated the inner branches like some grotesque candelabra. Ribbons of dry seaweed hung from shallow nests, all cemented together with copious amounts of the birds' own excreta. Even within the protection of the canopy, limbs and nests swayed unpredictably. Fragile chicks waited nervously for a parent returning with a crop full of fish or plankton, seemingly in imminent danger of eviction by the wind.

The Pelsaert colony at the time of our visit was estimated to number nearly 35,000 pairs, although populations appear to fluctuate considerably. In 1843, John Gilbert, assistant to the artist and naturalist John Gould, made the first trained observations of the Houtman Abrolhos. Gilbert's reports were filled with excitement for the vast numbers of birds, particularly Lesser Noddies. On Pelsaert Island he noted "amazing clouds [of birds] when congregating in the evening ..."[16] Around the turn of the century the colony suddenly disappeared, and in recent years their numbers on Pelsaert have varied from nearly 31,500 pairs in 1986 to more than 56,000 in 1989.

Huge fluctuations in seabird populations are quite common. In the high latitudes of the Houtman Abrolhos, the presence of Lesser Noddies and other essentially tropical species is dependent on the Leeuwin Current. Variations in the current's intensity and distance from the coast in any one season probably affect the productivity of surrounding waters and therefore breeding success.

Rock-lobsters Among the Sea-lions

The Leeuwin Current at the Houtman Abrolhos mingles with colder waters from the south, resulting in a unique mix of tropical and temperate marine species. Wondrously colourful corals compete with kelp for space and light, while lurid tropical fish cohabit with Australian Sea-lions (*Neophoca cinerea*) at the northern limit of their distribution.

Buffalo Bream (*Kyphosus sydneyanus* and *K. cornelli*), common fish on the Abrolhos reefs, graze and patrol curious polygonal territories of turf algae, "fenced" with boarders of kelp (brown macroalgae), all set amid an astonishing array of corals.

Despite fluctuations in the Leeuwin current and periodically lower water temperatures, tropical species are continually replenished by the southward flow. The Abrolhos reefs support more than 150 species of hard coral across 44 genera, a richness greater than that of the southern Great Barrier Reef.

I visited the Houtman Abrolhos in late November. Apart from my scientist companions from the Department of Conservation and Land Management (CALM) and the local Fisheries officer, all busily engaged in a seabird survey of the Pelsaert and Easter Groups — there was virtually no one around. Come the following February, everything would change. The annual rock-lobster season brings nearly 200 fishermen and their families to the archipelago. Twenty-two islands spread across the three groups are crowded with clusters of jetties and permanent shacks — homes away from home for three and a half months. From March until June, the fishermen work the Abrolhos waters, setting baited pots from a busy flotilla of boats, and trapping between them a season's catch of 1,000–1,500 tonnes of Western Rock-lobster (*Panulirus cygnus*). The Houtman Abrolhos also supports limited "wet-line" and scallop fisheries.

The Abrolhos lobster fishery is the richest in Australian waters — a lucrative, but strictly

controlled business. The season is timed to avoid conflict with the lobsters' breeding and migratory cycles, and catches outside its duration are illegal. Pot design and numbers are regulated by licence, as are the places they can be set. To ensure sustainability, all controls are vigorously enforced by the Fisheries Department of Western Australia.

However, commercial activity across each of the fisheries in the Houtman Abrolhos has not been without cost to the reef environment. Dragging anchors and some of the million or more pots dropped each year have significantly damaged the coral. Recreation has also taken its toll; many reef areas seem strangely empty of large fish, due to over-enthusiastic amateur spearfishing.

Overall management of the Houtman Abrolhos falls to the Fisheries Department while CALM is responsible for fauna protection on the islands. Both departments recognise the need for multiple-use zoning through a marine sanctuary system, a concept promoted since the mid 1970s. Unfortunately, a proposed Abrolhos National Park covering most of the islands, plus an associated marine reserve waits to be declared. A political rather than a management decision needs to be made.

Continuing along Pelsaert Island, I crossed to its spine of low and roughly parallel ridges of coral rubble, barely three metres above the surrounding sea. Cast up in ragged lines by the surf, dead-grey and bleached rosettes, florets and fans, crusty and hollow under foot, were all that remained from the exertion of generations of industrious coral polyps.

Exploring these fragments of the archipelago, I always felt the tension of history, the faintest echo of the brief but tumultuous events of three centuries ago. As if by bending down, I might find some tiny piece of zealous Dutch culture in the far flung Indian Ocean — a floral shard of Delft china among those dead aquatic flowers. While ever the reefs flourish and die, the Houtman Abrolhos will live up to its mysterious name.

A female Australian Sea-lion near her pupping site among mangroves on Suomi Island in the Easter Group. The Houtman Abrolhos is home to a small population of Australian Sea-lions—the only endemic seal found in Australian waters. Once far more abundant, they were an important food source for the Zeewijk survivors.

Western Rock-lobsters. Along the Western Australian coast, lobster larvae are carried far into the Indian Ocean by offshore surface currents soon after their summer hatching. At the mercy of the drift for a year, they then ride onshore currents encountered at depth, back to the continental shelf. Over the following four years the survivors grow to adulthood within the protected reef waters. Clay Bryce, Precision Images

(1) The term "abrolhos" was widely used by the Spanish and is close to their *abre ojos* also meaning "lookout" or "open your eyes". It has been suggested that the name "abrolhos" was applied directly to the Houtman Abrolhos by the Portuguese themselves but there is no direct evidence for this. A group of low-lying islands and shoals off the coast of Brazil also bear the name "Abrolhos" and may have inspired the naming of the Houtman Abrolhos.

(2) VOC: Vereenigde Oost-Indische Compagnie (Dutch East India Company).

(3) Named for Dirk Hartog's ship and the first recorded landing on the Western Australian coast in 1617.

(4) Named by J. L. Stokes, 1840. Stokes found a Dutch canon and other relics on the island. J. L. Stokes, *Discoveries in Australia, With An Account of the Coasts and Rivers Explored and Surveyed During the Voyage of HMS* Beagle *in the Years 1837–43*, T. and W. Boone, London, 1846 (facsimile edn Lib. Bd of SA, 1969) vol. 2, p. 149.

(5) The region receives an annual rainfall of about 470 millimetres, most of it resulting from winter storms.

(6) Randolph Stow, *The Merry Go Round in the Sea*, [1958], Picador, Sydney, 1983, p. 113.

(7) F. de Houtman quoted in H. Edwards, *The Wreck on the Half Moon Reef*, Rigby, Adelaide, 1970, pp.82-3. De Houtman is said to have noted the hazard by marking "abrolhos" on his chart, hence the connection of his name with the Portuguese corruption.

(8) Commanded by Jan Pieterzoon Coen, the great Governor-General of Batavia and architect of Dutch power in the East Indies, it was saved only by chancing upon the Houtman Abrolhos in daylight.

(9) Two miraculous jewels consigned for sale in the East were wrested intact from the mutineers. One, a vase carved from solid agate, was the property of the great artist and diplomat Peter Paul Reubens. The *Batavia* was also carrying 137 worked sandstone blocks intended for erection as a portico façade on Batavia Castle. The assembled blocks are some of many artifacts raised from the wreck and now displayed along with a portion of the *Batavia*'s stern at the Western Australian Maritime Museum in Fremantle.

(10) "... On the south-west point of the island the beams of a large vessel were discovered, and as the crew of the Zeewyk ... reported having seen the wreck of a ship on this part there is little doubt that the remains were those of the Batavia." Stokes, *op. cit.*, p. 138.

With the translation of Pelsaert's journals at the end of the nineteenth century doubts arose as to the match of his descriptions of the backdrop to the tragedy and the islands of the Pelsaert Group. Brilliant detective work by the author Henrietta Drake-Brockman led to the discovery of the *Batavia* in 1963. See H. Drake-Brockman, *Voyage to Disaster, The Life of Francisco Pelsaert*, Angus and Robertson, Sydney, 1963.

(11) See P. J. Fuller, A. A. Burbidge and R. Owens, 'Breeding Seabirds of the Houtman Abrolhos, Western Australia: 1991–1993', *Corella*, 1994, 18(4): 97–113.

(12) The origins of the name are obscure but it is believed that sailors may have once called them "noddies" or "noodles" (simpletons) because of the ease with which they could be caught.

(13) Stokes, *op. cit.*, p. 146.

(14) The Pelsaert Island deposits were even mined during World War II when chemical fertilisers were in short supply.

(15) Populations of the other subspecies of Lesser Noddy (*Anous tenuirostris tenuirostris*) are found in the western Indian Ocean, centred on the Seychelles.

(16) H. M. Whittell, 'A review of the work of John Gilbert in Western Australia', *Emu* 41, 1942.

HOUTMAN ABROLHOS

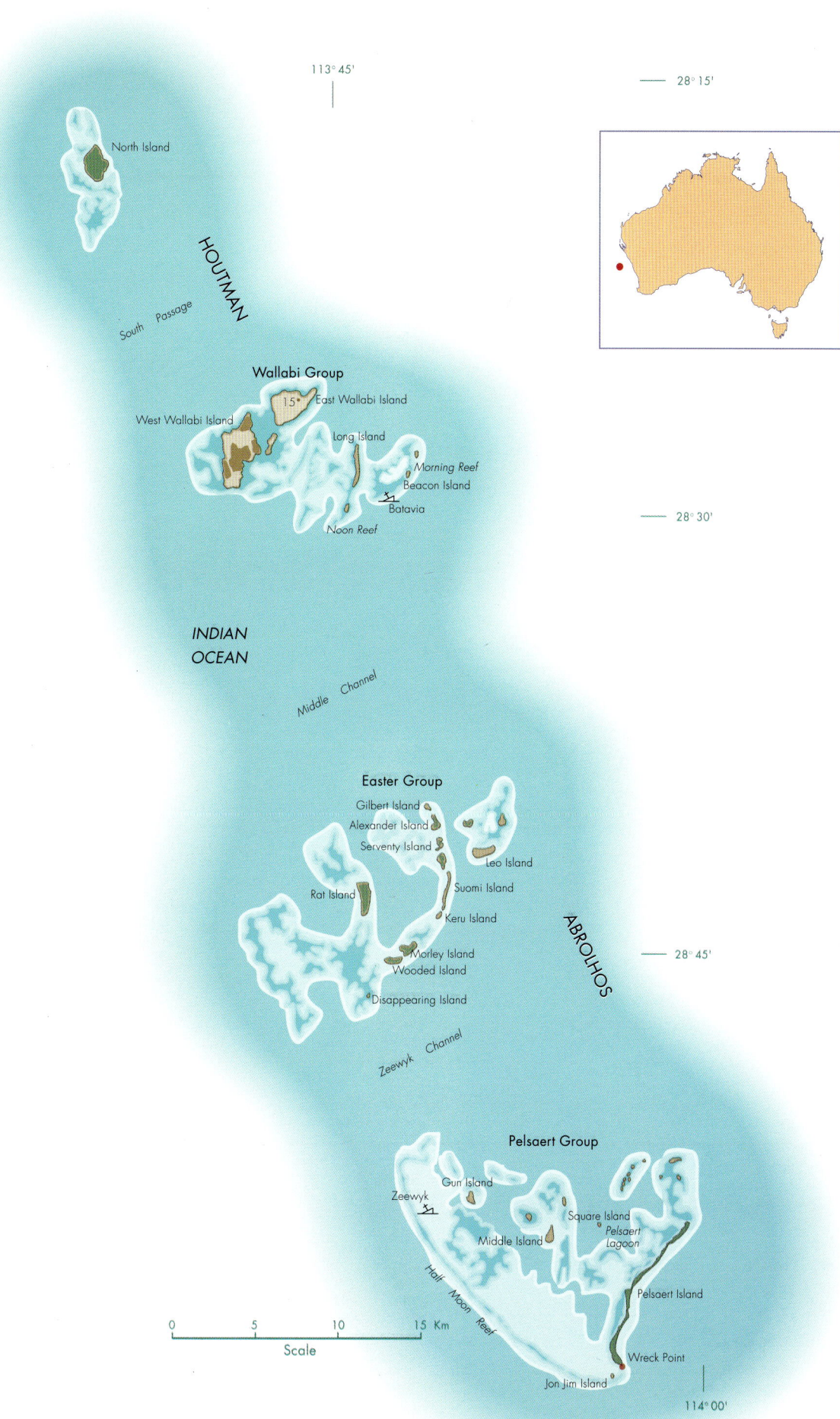

HOUTMAN

South Passage

North Island

Wallabi Group

West Wallabi Island

East Wallabi Island

15°

Long Island

Morning Reef

Beacon Island

Batavia

Noon Reef

**INDIAN
OCEAN**

Middle Channel

Easter Group

Gilbert Island

Alexander Island

Serventy Island

Leo Island

Rat Island

Suomi Island

Keru Island

Morley Island

Wooded Island

Disappearing Island

Zeewyk Channel

ABROLHOS

Pelsaert Group

Gun Island

Zeewyk

Square Island

Middle Island

Pelsaert Lagoon

Half Moon Reef

Pelsaert Island

Wreck Point

Jon Jim Island

113° 45'

28° 15'

28° 30'

28° 45'

114° 00'

0   5   10   15 Km

Scale

LOCATION

28°18'S to 29°00'S, 113°35'E to 114°02'E.
About 70 km off the WA coast. More than 120 islands, islets and rocks in 3 main groups. Geraldton is the closest town.

AREA

West Wallabi Island: 600 ha
Pelsaert Island: 120 ha
Rat Island: 57 ha
Gun Island: 3 ha

CLIMATE

Mediterranean, windy

STATUS

All islands are Class A Reserves for 'Conservation of Flora and Fauna, Tourism and Purposes associated with the Fishing Industry'. Abrolhos Islands National Park is planned. Managed by the Fisheries Department of Western Australia. Department of Conservation and Land Management (CALM) is responsible for conservation of the fauna.

ACCESS

Occasional private or charter visits for diving, bird watching etc. Yacht access difficult. Permits required. Consult Fisheries and CALM.

FACILITIES

No public facilities.

BERNIER AND DORRE ISLANDS

"... lying back in the easy chair of the sea"[1]

An embracing calm settled over Red Cliff Point and the waters of Shark Bay. Arrayed below, a line of cormorants flapped determinedly towards their roost for the night approaching, back to the security of some oft-used inshore rock-stack. The day still refused to die. Not until the shadow of Bernier Island itself dissolved in the rose and purple after-light did Shark Bay finally lay down with the ease of the night.

Bernier Island, and Dorre, its near neighbour to the south, two roughly cigar-shaped remnants of a drowned coastline, sit point to point at the northern end of Shark Bay within 50 to 60 kilometres of the coast of Western Australia at Carnarvon. Free of human habitation for all but a tragic moment in their being, both have become priceless in this era of accelerating animal extinctions, largely due to their isolation.

Long, thin and detached — like the ruins of some geological seawall built to guard Shark Bay — they seem not truly part of it compared to the much larger Dirk Hartog Island, 25 kilometres to the south of Dorre Island's Cape St Cricq. Dirk Hartog Island thrusts north-west — a disembodied arm, worked free of the intricate geography of "prongs" and "loops" that make up the bay side of Edel Land, the coastal peninsula of Shark Bay.

Because of its geographic configuration and arid surrounds, much of Shark Bay is not flushed by the open sea or inflowing rivers. Increased salinity (up to double that of the open ocean) restricts the development of coral throughout the bay, except for the inner shores and ocean passages of southward-lying Dirk Hartog Island, as well as Bernier and Dorre Islands. There are no extensive reefs but around the gap between Bernier and Dorre, shallow rocky platforms support diverse gardens of staghorn coral (*Acrophora* spp.).

Rich in marine life, Shark Bay is home to turtles, dolphins, sharks, seasnakes and a multitude of fish. An estimated 10,000 Dugongs (*Dugong dugon*) graze on more than 1,000 square kilometres of seagrass meadow. Humpback Whales (*Megaptera novaeangliae*) loiter on their migration along the coast. Shark Bay is famous for its population of Bottle-nosed Dolphins (*Tursiops truncatus*) while Killer (*Orcinus orca*), Pilot (*Globicephala* sp.) and Pygmy Sperm Whales (*Kogia breviceps*) have all been sighted within its reaches.

Pleistocene Dunes

Both islands are composed of limestone which accumulated as enormous sand dunes on the region's western shoreline during the extremely dry glacial periods of the Pleistocene Epoch. About two million years ago incessant southerly gales began to sweep up dunes that reached their zenith to the south of Dirk Hartog Island. Here the 200 kilometre-long Zuytdorp Cliffs rise in places to more than 250 metres.

During the last million years, sea level changes associated with world glacial fluctuation have seen Shark Bay flooded and drained repeatedly. One of these flooding episodes carved off a sliver of coastline about 8,000 years ago. Stabilising at current levels 4,000 years later, Shark Bay was by then close to its present configuration. As recently as 3,000–5,000 years past, surf pounding on the western shores of the newly parted fragment carved a one kilometre-wide channel through to Shark Bay, finally separating Bernier and Dorre Islands from one another.

Overlying the lime-cemented rock of both islands, the most recently accumulated calcerous sands form extensive sandplains and dune fields. Significant numbers of these dunes are unconsolidated and remain mobile, particularly on Bernier Island. An exposed travertine limestone pavement runs for nearly the entire western cliff-edge of both islands where sand has been stripped by the wind from the underlying rock.

Opposite: The yacht Bydand *rides gently at anchor just north of Red Cliff Point on Bernier Island. Such tranquillity can be shortlived, with neither Bernier nor Dorre Island offering any shelter from frequent, strong southerly and south-easterly winds.*

Cape Boullanger near Disaster Cove on Dorre Island, looking north towards Bernier Island. It was here in 1839 that explorer George Grey and a party of 11 men were almost swept away in a gale. Suffering the extremes of Shark Bay weather, Grey wrote that at the height of the storm they were "benumbed and miserable with cold", only to be "oppressed with thirst ... [upon] the coming of another roasting day."

Shifting Sands

The raw power of wind and wave, still moulding the breach between the two islands, is dramatically recounted by the hapless explorer, and later Governor of South Australia and New Zealand, George Grey. After sailing from Fremantle, Grey was dropped off near Bernier Island in February 1839 with a party of 11 men and three whale boats. His plan was to investigate the coast and hinterland from Shark Bay to North West Cape. On 28 February, while the party was at the northern end of Dorre Island, a great storm struck.

"We all quailed or fell before it, for it came with sudden and indescribable violence ... the bushes were stripped from the ground [and] the sea gained on us so much that I had made up my mind it would sweep away the intervening sand hills, and once more wash the face of the cliffs. In this case, we should, to a certainty, have all perished ... [We] were benumbed and miserable with cold."[2]

Strong southerly and south-easterly winds are common on Shark Bay, and the islands as witnessed by Grey, afford no protection in storms. In general though, the bay enjoys a dry, warm, Mediterranean climate as could be expected for this part of Western Australia. Carnarvon, on the coastal fringe of the arid Gascoyne region, receives an annual average rainfall of just 219 millimetres, most of which falls between May and August. Summers are hot, with average January maxima in parts of the region pushing towards 40° Celsius. There is no surface water on either island.

Their flora very much reflects the semi-arid climate. Lying at the boundary of two botanical divisions, desert species mix with those from the south-west of the State at the northern limit of their range. Sandplains occupy the greatest area on both islands but the mix of species for each are distinctively different. Much of Dorre Island's sandplain is covered by hummock grassland, dominated by the spinifex *Triodia plurinervata* (a species endemic to Shark Bay), interspersed with dwarf shrubs such as *Thryptomene micrantha*. On Bernier Island the sandplain is covered by shrubs rather than grassland, principally *Abutilon exonemum*, *Scaevola crassifolia* and *Acacia* species. Consolidated dunes on both islands bear shrubs such as Sandalwood (*Santalum spicatum*), *Diplolaena dampieri* and *Acacia coriacea*. In dune swales on Dorre Island the scrub can reach two metres in height. Restless dunes of beach sand are more sparsely vegetated. Salt-pruned colonising plants creep after drifts of shifting sand, their roots often scoured bare by the wind. More common on Bernier Island, these unconsolidated dunes are dominated by *Pilaenthus limacis*, a species endemic to Shark Bay.

Creatures of the Night

It was almost dark when I returned to my camp pitched between spinifex and tide, not far north of Red Cliff Point, on Bernier Island. Newly awakened stars brightened in a deep indigo sky. A few hundred metres offshore, the masthead light of our yacht traced out a short arc as it swayed back and forth among the firmament.

I set my mind to the business of food. A delicious hint of freshness came to the air as evening dew began to settle over sand and spinifex. This was my first evening on the island. Thoughts of the hours ahead, stumbling across the sand hills in search of the creatures of the night made me slightly apprehensive. But I need not have worried. As if on cue, the night's activities commenced.

A dune spills onto the beach at Red Cliff Point, Bernier Island. The extent of unconsolidated dunes has been greatly exacerbated by feral goats. Persisting on Bernier Island for more than 50 years, they were finally eradicated in the 1980s.

The Burrowing Bettong or Boodie is unique among Macrapods for its year-round burrowing habit. By day Boodies live in extensive communal burrows, only emerging to forage after dark. With their range so drastically reduced to just four islands off Western Australia, efforts are being made to reintroduce them to fenced areas on the adjacent mainland, cleared of feral predators and competitors.

Opposite: Dunes consolidated as limestone in the Pleistocene Epoch crumble and erode in fanciful formations on the east coast of Bernier Island near Red Cliff Point.

Behind the cover of grass, something twitched. Then silence. Again the grass rustled as I caught a glimpse of an animal about the size of a rabbit hopping between cover. Soon two animals ventured forth. This time they paused, sat back on their haunches and sniffed the air with interest. Obviously attracted by the food I was cooking, more animals emerged until I was confronted by a group of up to a dozen hopping, squabbling marsupials. Their timidity had disappeared — as the aroma of my cooking grew stronger so their approaches became bolder. My pugnacious companions were Burrowing Bettongs or Boodies (*Bettongia lesueur*).

They were once one of Australia's most abundant mammals, inhabiting extensive warrens throughout much of arid and semi-arid Australia as far east as the Great Dividing Range. But their subterranean homes also proved popular with invading rabbits. Early accounts describe the sociable Boodies cohabiting with rabbits for decades following the invasion, but inevitably the Boodies went into decline as competition for food and living space increased. As well as rabbits, predation by foxes and cats, grazing and trampling by domestic stock, and the change from Aboriginal to European land management methods, all hastened the Boodies' decline. They were probably gone from the mainland altogether in the 1940s, restricted by then to a few Western Australian islands comprising less than one percent of their former range. Cats and trampling sheep rendered them extinct on Dirk Hartog Island, leaving only four islands — Barrow, Boodie, Bernier and Dorre — as their entire remaining habitat.

During my stay on the two islands, Boodies maintained an ever hopeful vigil around my camp each night. It soon became almost impossible to prepare food in their presence as they would pounce on anything vaguely edible that was not actually in my clutches. They even mounted determined assaults on my stove and its scalding contents. Somewhat intimidated by these diminutive assailants, I was forced to do most of my food preparation before they emerged from their burrows for an evening's marauding.

Just as Boodies became my audacious

companions, so other creatures of the night kept their cover, timid and silent, venturing forth to feed in the cool veil of darkness. The Shark Bay Mouse (*Pseudomys fieldi*), a secretive inhabitant of the beach grasses of Bernier Island proved elusive. Bernier is now the only stronghold for this tiny mammal, which was once found on the mainland fringes of the bay. Weighing less than 50 grams, it nests in the dunes among clumps of Beach Spinifex (*Spinifex longifolius*).

Amid the antics of the Boodies, I occasionally caught a fleeting glimpse of another small mammal — the Western Barred Bandicoot (*Perameles bougainville*) — as it nervously scurried among the grass and shrubs of the beachfront. Like the Boodie, the Western Barred Bandicoot has also suffered a catastrophic decline. Bernier and Dorre Islands are its last refuge. With a pre-European distribution similar to, though not as extensive as the Boodie, its decline can be attributed to many of the same sad factors.

Walking along the beachfront, I encountered numerous pale yellow crabs scurrying for the cover of their sandy lairs. I struck inland from the beach keen to catch a glimpse of the less obvious creatures of the night. Out in the sandhills the air was filled with the boom of the surf from the seaward coast close by. There were animal tracks everywhere. No discernible pattern or purpose — short dashes or broad meanderings — they all spoke of much activity on this and several nights past.

Despite such evidence, it was some time before my torch-beam fell on a small wallaby nibbling the tender shoots of Beach Spinifex. Dazzled by the unaccustomed glare, it froze, staring in the direction of the light, large liquid eyes glowing red. I moved slightly and it was off, vanishing among the dunes. It was a Rufous Hare-wallaby or Mala (*Lagorchestes hirsutus*), one of two species of wallaby found on the islands. As my stay progressed, I saw many Rufous Hare-wallabies quietly grazing each night among the foredunes near Red Cliff Point. During the day they shelter in shallow burrows and scrapes, frequently under hummocks of spinifex or among low heath.

Just as common is the Banded Hare-wallaby (*Lagostrophus fasciatus*), an habitué of the low thickets of *Acacia coriacea* and other dense shrubs of Dorre's inland consolidated dunes and west coast travertine pavements. They are also locally abundant among dense heath at the northern end of Bernier Island.

Like the other mammals of Bernier and Dorre Islands both species of wallaby have been reduced to relict island populations.[3] Once an important game animal for Aborigines in Central Australia, the Rufous Hare-wallaby had disappeared from the better watered parts of its range by early this century, but lingered in the northern deserts of Western Australia, possibly until about 1960.

The Banded Hare-wallaby has vanished altogether from its mainland range in the south-west of Western Australia, not being sighted since 1906. Bernier and Dorre Islands are its last stand. Less than 4,000 individuals survive on each island.

"... a sort of racoons ..."

The animals of Shark Bay were among the first to be collected and described by European explorers. In 1699, William Dampier aboard the *Roebuck* on his second voyage to New Holland, coasted into a broad sound, noting: "The sea fish that we saw here are chiefly Sharks ... I therefore give it the Name of Shark's Bay."[4]

He landed near Cape Levillian at the northern end of Dirk Hartog Island and made the first scientific observations of the continent and its natural history. Of the mammals he wrote: *"The Land-Animals that we saw were only a Sort of Racoons, different from those of the West-Indies, chiefly as to their Legs; for these have very short Fore-Legs; but go jumping upon them as the others do."*[5]

Dampier's frustratingly quaint description gives no clues as to the animal species he encountered, but a series of French visits to Shark Bay in the late eighteenth and early nineteenth centuries provided more decipherable observations of the natural features. The first, in 1772, was by Saint Allouarn, followed in 1801 by Nicolas Baudin's *Le Géographe* Expedition. *Le Géographe* returned in 1803. Baudin's visits to Shark Bay were among the first, consistent attempts at scientific exploration, collection and observation in Australia. Yet Baudin remained unmoved by the landscape, describing the cliffs of Dirk Hartog as sterile and uninviting, "like the lands of Sinai". He called Bernier and Dorre Islands the "barren islands". On its 1801 visit, the expedition established an astronomical observatory on Bernier Island, which was named for Pierre François Bernier, astronomer on *Le Géographe*. Meanwhile, naturalist François Peron studied the wildlife, observing the "Striped Kangaroo" in great numbers on the three main islands of the bay. Commenting on their vulnerability, he found the Banded Hare-wallaby "unprovided with the means of attack or defence ... mild and timid"[6].

Threats and Security

The wildlife of Bernier and Dorre has survived since its separation from the mainland. Through to the era of exploration and beyond, the catastrophe of feral predators that befell so many islands has not occurred. However, Bernier and Dorre have been disturbed. Thankfully, sheep-grazing at the turn of the century failed, but a feral goat infestation persisted on Bernier Island, being shot out only in 1981. Large stretches of unconsolidated dune remain, the result of erosion precipitated by ravenous goats. Sandalwood was cut on Bernier Island in the 1890s and fire has swept Dorre Island three times. The last of these fires, in 1973, destroyed much of the Banded Hare-wallaby's habitat. Return of vegetative cover has been remarkably slow and it may never regain its original state.

In recognition of their critical conservation significance, both islands were gazetted in 1957 as nature reserves for the preservation of their mammal fauna. Because of the potentially disastrous consequences of further fires or the introduction of exotic species, public access is restricted.

"...tombs of the living dead"

The peace I drank in each evening while on the islands, and the sanctuary afforded their animal residents is hard to reconcile with the anguished description of Bernier and Dorre given in 1938 by Daisy Bates, that indomitable advocate for Aboriginal welfare. Bates spent several months on the island in 1910–1911 in her capacity as Western Australia's government attaché to a Cambridge University anthropological expedition. She wrote of great suffering:

"...there is not, in all my sad sojourn among the last sad people of the primitive Australian race, a memory one-half so tragic or so harrowing ... as these two grim and barren islands of West Australian coast that for a period, mercifully brief, were the tombs of the living dead."[7]

A Rufous Hare-wallaby or Mala grazing at night on Dorre Island. Recent surveys estimate that there are just 2,600 on Bernier and about 1,700 on Dorre Island. These are the last remaining wild populations of this species, once widespread throughout much of arid and semi-arid Australia.

A typical shelter provided for Aboriginal male "patients" of the lock hospital at the northern end of Bernier Island, c. 1910. After brutal state government measures outlawing cohabitation between different races, Aborigines suffering from venereal diseases were taken in chains to Carnarvon and from there to the islands. Sick, disoriented and emaciated, the men were unable to hunt or fish. Despondency prevailed. The women on Dorre Island suffered similarly. The hospitals remained open for nine years. Courtesy of Battye Library, Perth

In one of the most deplorable episodes in the history of treatment meted out to Aborigines, Bernier and Dorre Islands were used between 1908 and 1917 as isolation or "lock" hospitals for tribespeople suffering from venereal diseases.

The "patients" (more accurately, prisoners) were rounded up from all parts of north and north-east of Western Australia. "Regardless of tribe and custom and country and relationship," Bates noted, "they were herded together — the women on Dorre and the men on Bernier. Many had never seen the sea before and lived and died in terror of it."[8]

Treatments were crude and the hospitals inadequately staffed. The patients were put to work despite their condition. The men laboured on the buildings, erected fences and tended domestic stock, while the women cooked, and collected firewood and water.

Abandoned in 1917 for shore-based treatment of the survivors at Port Headland, the lock hospitals accommodated over 700 Aborigines in their time. Records were poorly kept, particularly in the early years, but it is likely that more than 200 people died on the islands or in transit. Those that received a "discharge" during the life of the hospitals were, in reality, abandoned. Dumped at Carnarvon, they were left to find their own long way home, often across the territory of hostile tribes. Their fate remains unknown.

Scraps of Land — Arks of Life

Another gentle evening descended over the cliff-tops and dunes of Bernier Island. Unruly lines of terns winged past to unknown beaches and colonies among the crumbling limestone bluffs to the south. Out of sight to the east a whole continent rolled into darkness as the last of the day's sunlight brushed the dunes. At that moment, the haunting, lonely glow of evening seemed only to cast into relief the sadness and misery that was for a time Bernier and Dorre Islands.

I had been camped on Bernier Island for nearly a week. Each day as I roamed its dunes, thrilled by the record of life scribbled in the sand, the priceless importance of such places had been made abundantly clear. As so many unique, small and timid mammals have been expelled through silent extermination from the content that spawned them, it is truly providential that an array of original species can still be found on those two scraps of land between such a great bay and the open sea.

(1) J. Olsen, M. Durack, M. Serventy, G. Dutton, and A. Brotignon, *The Land Beyond Time*, Macmillan, Melbourne, 1984.

(2) G. Grey, *Journals of Two Expeditions of Discovery in North-West of Western Australia 1837–1839*, T. and W. Boone, London, 1841 (Facsimile edn, Hesperian Press, Perth, 1983), vol. 2, p. 342.

(3) The Rufous Hare-wallaby survived until recently in two tiny populations in the Tanami Desert. Attempts are being made by the Parks and Wildlife Commission of the Northern Territory to reintroduce it to parts of its Tanami range in conjunction with cat and fox control measures.

(4) W. Dampier, *A Voyage to New Holland*, London, 1729 (J. Spencer [ed.], edn Allan Sutton, Gloucester, 1981), p. 110.

(5) *Ibid.*, p. 109.

(6) quoted from Baudin's journal in R. Radok, *Capes and Captains*, Surrey Beattie, Sydney, 1990, p. 215.

(7) D. Bates, *The Passing of the Aborigines*, Heinemann, Melbourne, 1966, pp. 96–7.

(8) *Ibid.*, pp. 97–9.
     Curiously, in her report of the time to the government, Bates remarked favourably on the condition of the inmates, revising her opinion in later years, heavy with sadness "upon the ghastly experiment of Dorre and Bernier Islands". *Ibid.*, p. 108.

BERNIER AND DORRE ISLANDS

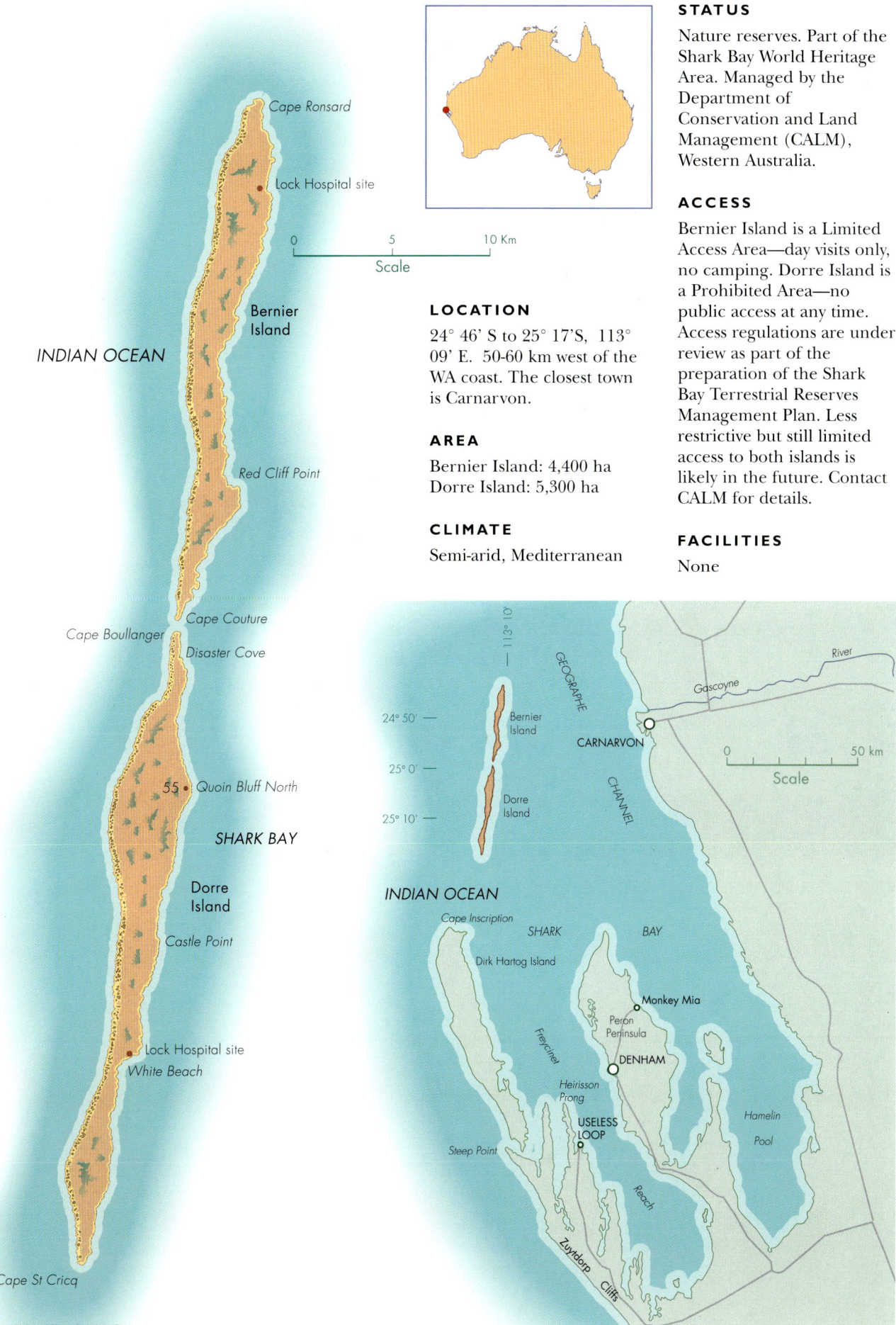

STATUS

Nature reserves. Part of the Shark Bay World Heritage Area. Managed by the Department of Conservation and Land Management (CALM), Western Australia.

ACCESS

Bernier Island is a Limited Access Area—day visits only, no camping. Dorre Island is a Prohibited Area—no public access at any time. Access regulations are under review as part of the preparation of the Shark Bay Terrestrial Reserves Management Plan. Less restrictive but still limited access to both islands is likely in the future. Contact CALM for details.

FACILITIES

None

LOCATION

24° 46' S to 25° 17'S, 113° 09' E. 50-60 km west of the WA coast. The closest town is Carnarvon.

AREA

Bernier Island: 4,400 ha
Dorre Island: 5,300 ha

CLIMATE

Semi-arid, Mediterranean

Bean Blossoms

Coral mushrooms wavered in the glassy refraction of the dinghy's bow wave as we skimmed towards Malus Island, our first landfall in the Dampier Archipelago. Approaching the shimmering white beach next to Courtenay Head, splotches of scarlet could be seen sprawling across the strand. They were runners of the distinctive Sturt's Desert Pea (*Clianthus formosus*), looking oddly out of place so close to such vivid aquamarine waters. But this cluster of islands, hard by the coast near the mining port of Dampier, is very much the maritime expression of the great semi-desert Pilbara region.

Like the mainland Pilbara, they are old — an age of truly archaic proportions. And just as this ancient region has now been bent to the needs of an insatiable modern world, so the archipelago forms a mute backdrop to a twentieth-century industrial parade.

Sturt's Pea and other dune plants of these islands were among the first botanical specimens to be collected for science in Australia. In September 1699, William Dampier in command of the *Roebuck*, made the earliest known European landing in the archipelago during his second voyage to New Holland.[1]

"[Dampier] found 2 Sorts of Grain like Beans: The one grew on Bushes; the other on a Sort of a creeping Vine that runs along the Ground, having very thick broad leaves and the Blossom like a Bean Blossom, but much larger, and of a deep red Colour, looking very beautiful."[2]

"... the Stones were all of rusty Colour . . ."

In addition to his cryptic observations of flora, Dampier gave scant descriptions of the landforms themselves:

"...all appeared dry, and mostly rocky and barren ...

but [I] was in Hopes of finding a Channel to run in beyond ... that amongst so many Islands we might have found some Sort of rich Mineral, or Ambergreece, it being a good Latitude for both of these ... The Stones were all of rusty Colour, and ponderous."[3]

Preoccupied with finding water, Dampier quit the islands after a brief and unsuccessful search. His "ponderous stones", the dramatic rockpiles of the archipelago are among the oldest rocks on earth. They are fragments of the Pilbara Block, where dates of more than 3,500 million years have been recorded. Outcrops of Archaean granite and granite gneiss on Dolphin, Enderby and Tozer Islands are at least 2,800 million years old. In contrast, the Collier Rocks and Delambre, Hauy and Keast, among the smaller islands, are composed of very recent calcerous dune limestone that accumulated in the Pleistocene Epoch. Embellishing the margins are sandplains, muddy mangrove flats and beaches formed in the last 10,000 years. Only in the last 8,000 years has the archipelago itself been freed by rising seas from the tether of the mainland.

Opposite: More than 2,500 million years separates the geologically recent limestone mushrooms of Collier Rocks from the ancient granite rockpiles of Dolphin Island, a kilometre or so across Flying Foam Passage.

Left: An opportunist, Sturt's Desert Pea takes advantage of erratic deluges from summer cyclones and tropical depressions that bring the Dampier Archipelago nearly all its meagre annual rainfall of about 275 mm. After a soaking, Sturt's Peas and other annuals germinate rapidly. Sprawling over the dunes, they flower and seed quickly before dying in the baking tropical heat.

Cheek by Jowl

Courtenay Head's prominent bluff affords a commanding view in early every direction. Islands are scattered about, tethered like ships in the stream. A few kilometres beyond, illusion turns to reality: I counted at least 20 sheer-sided bulk carriers, riding at anchor on the horizon. Each was waiting its turn to ply Mermaid Sound and tie up at Dampier's wharfs. There they would engorge salt, liquefied natural gas or a cargo of iron ore from the open-cut mines of the Pilbara hinterland.

Back over the rise, an open-webbed tower pushes skywards. This gargantuan structure standing in mid-current off Malus Island is just one component of the network of offshore gas platforms, submarine pipelines, port facilities and industrial infrastructure that comprises Woodside Offshore Petroleum's North West Shelf project, Australia's largest resource development. The Dampier Archipelago lies at its hub.

As the day began to fade, I walked back along the cliff-edge to Marney Bay. The most unlikely of "seabirds", the Little Corella (*Cacatua pastinator*), crowded in a garrulous gang, about 50-strong, on a tumbled band of rocks below. Dampier had observed these birds of the arid interior, noting "a sort of white Parrots, which flew a great many together".[4] Consistent with the archipelago's semi-desert character, Little Corellas have been recorded breeding on Malus and East Intercourse Islands. Above their rasping cries, I heard another call: the thin, high-pitched whinnying of an Osprey (*Pandion haliaetus*).

It was late June and the breeding season had been in progress for nearly two months. Osprey pairs were sharing incubation duties or

Opposite, top: An Osprey lands on its nest stack. Found around much of the Australian coast, Ospreys are primarily fish hunters. Diving feet-first, they snatch their prey from just below the surface.

Opposite, below: Dense hummocks of the spinifex Triodia wiseana *growing among rockpiles at the northern end of Dolphin Island. Legendre Island lies in the distance. Legendre remains outside the Dampier Archipelago Nature Reserves, set aside for the future development of an industrial port to service expansion of the North West Shelf project.*

Left: The boulder-strewn slopes of Dolphin and other large islands support a sparse scattering of shrubs less than two metres high. These include Acacia pyrifolia, Grevillia pyramidalis *and* Hakea suberea *(pictured).*

building and repairing their flamboyant nest stacks. Woven from sticks, grass, seaweed and seemingly any flotsam, the nests are used year after year and in time reach substantial proportions — sometimes nearing the size of a small car. Ospreys live year-round in the archipelago and are one of 15 sea and shore species known to breed there.

I moved on. Pausing to look back, I noticed that one bird had landed with fresh nesting material, though its mate still hovered nervously. The nearby tower formed a curious backdrop to this wild view. From casual observation, industry and nature were co-existing cheek by jowl in some degree of harmony — or so it seemed.

The Intercourse Islands

For thousands of years before Woodside began searching for submerged wealth off the Pilbara coast, small groups of Aboriginal people lived here, exploiting the living wealth of land and sea.

Although Dampier observed the local Aborigines, the earliest extensive European observations of them were made by Phillip Parker King in early 1818.[5] Attesting to widespread activity, King noted: "*The tracks of natives and their fire-places were every where visible, and around the latter the bones of kangaroos and fishes were strewed ...*" The next day, Aborigines were encountered fishing:

Petroglyphs, Dolphin Island. The granophyre boulders of the Dampier Archipelago and the Burrup Peninsula are the repository for one of Australia's greatest rock art galleries. More than 10,000 engravings are spread across 500 sites on the peninsula alone. They are the only reminder of the Yapurrarra people, a coastal group who failed to withstand the advent of Europeans in the Pilbara.

"It appears that the only vehicle, by which these savages transport their families and chattels ... was made of a stem of a mangrove tree; but as it was not long enough for the purpose, two or three short logs were neatly and even curiously joined together end to end, and so formed one piece that was sufficient to carry and buoyant enough to support the weight of two people."[6]

The Aborigines observed by King were most likely Yapurrara, the traditional people of the islands and nearby Burrup Peninsula. The Yapurrara were probably active in the region from about 18,000 years ago, roaming the coastal plains and peninsulas for generations,

before inundating seas turned the higher hills to islands. Undeterred by rising waters, these maritime people travelled on log rafts, making seasonal trips to the islands, where they feasted on the abundant marine life thereabouts: rock oysters, fish, turtles, turtle eggs and Dugong. Lack of freshwater limited the timing and duration of their visits. Water flowed only during or immediately after rain, forcing the Yapurrara to rely on ephemeral pools or shallow wells dug in the sand.

Flying Foam and Tranquillity

Early afternoon, and the wind backed from the consistent and invariably strong easterlies to a pleasant north-west sea breeze. It was time to leave Malus Island and head east. We weighed anchor and sailed on the wind for the southern end of Flying Foam Passage and a tight cluster of islands about 12 kilometres away. Turning north into the quiet waters between Angel, Gidley and Dolphin Islands, such a peaceful tropical waterway did not seem to fit with such a descriptive name. The *Flying Foam* was a nineteenth-century pearling lugger whose colourful name shadows an era of great brutality in Western Australia's history.

The pastoral industry and gold mining brought enormous disruption to Aboriginal life in the north-west, but neither equalled the horrors that came with pearling. Pearl shells were found at Shark Bay in 1850, and by 1860 a pearling fleet operated from Cossack on the Pilbara coast. From about 1870 to 1900, Flying Foam Passage was the major pearling ground on the north-west coast. In the gross cruelty that characterised these 30 years, the Aborigines of the region were decimated, the Yapurrara among them. On a lawless frontier, slavery, rape, disease and deprivation ruled. The Reverend J. B. Gribble, missionary and fierce defender of Aboriginal rights in the north-west, wrote in 1886, quoting a first-hand source:

"It is well known ... that in one day [in 1868] there were quite sixty natives, men women and children, shot dead ... I saw at 'Flying Foam Passage' no less than 24 natives handcuffed together, and then

conveyed to Delambre Island, and there detained, until they were required for pearl diving, their only food being a little flour."[7]

Either murder, disease or an agonising death from the bends accounted for all the Yapurrara. Of an original group of fewer than 120, none survived into this century.

Rothschild's Among the Rockpiles

The islands flanking our progress rose in steeply rounded forms, straight out of turquoise waters to an undulating, treeless skyline. Slices of white sand drew a broad line on the tide as piles of deep-russet granophyre boulders appeared to tumble from ridges amidst swathes of gold and green desert herbage. At the northern end of Dolphin Island we came to anchor in a shimmering afternoon calm, broken only by the blow of a pair of dolphins negotiating the inshore passage.

Several species of marine mammal cruise the archipelago. Humpback Whales (*Megaptera novaeangliae*) with newly born calves loiter around the islands from June to September as they slowly migrate south for the summer. In the 1870s, this migration was often abruptly ended by whalers operating from a shore station and processing works on Malus Island.

With an area of 3,203 hectares, Dolphin is the second largest of the 42 islands, islets and rocks that make up the 45-kilometre sweep of the Dampier Archipelago.[8] It is also the highest, rising to 120 metres along its central spine. With this in mind, I set out the following morning for the island's summit.

The best route up looked to be via one of the many steep but unvegetated rockpiles. Barely past the tidal skirt of scattered mangroves and mud-smeared boulders, I came across an enduring reminder of the island's lost people: the top surface of a roughly rectangular slab was pecked out with images of fish, sting-rays and an assortment of indecipherable symbols. Nearby, a large upright rock sported a lone turtle.

The going was deceptive. The bare rockpiles

acted as heat-sinks, making climbing hot work in the reradiated late morning sun. As I clambered upwards, huge boulders unexpectedly teetered on hidden fulcrums, stone striking stone, tonnes of rock ringing like an anvil.

Despite fierce surface conditions, the rockpiles are important refuges for a variety of animals. Deep slots up to 15° Celsius cooler than the outside air offer relief on the larger islands for the Euro (*Macropus robustus*) and the Rothschild's Rock-wallaby (*Petrogale rothschildi*).[9] Similarly, the Little Northern Native Cat or Northern Quoll (*Dasyurus hallucatus*) and Common Rock-rat (*Zyzomys argurus*) shelter in the rocky shade on Dolphin Island, venturing out to hunt or forage in the cool of night on the adjacent sandplains. These mammals share the rockpiles with the rare Pilbara Olive Python (*Morelia olivacaea barroni*), one of more than 40 species in the archipelago's diverse reptilian fauna. From the secrecy of deep crevices, Olive Pythons ambush small mammals, birds and lizards.

Honeyeaters were active among the shrubs fringing the boulders. Burbling with song, they called back and forth across the saddle, their voices like bells, chiming on iron-hard rock. A small golden-breasted bird, possibly a Grey-headed Honeyeater (*Lichenostomus keartlandi*) flitted from bush to bush but concentrated most of its attention on a lanky *Hakea suberea*, drooping under the weight of its pendulous yellow flower spikes, each dripping with nectar.

Pearling luggers at low tide on the Flying Foam Passage, c. 1900. In the 1870s about 500 people, including Europeans, Malays, Chinese and Aborigines, worked the passage's pearling fleet. The fleet remained the region's largest until about 1900. Courtesy of Battye Library, Perth

33

*With its nest nearby, a Caspian Tern (*Hydroprogne caspia*) calls a warning. The smaller islets and rocks of the archipelago are important breeding sites for sea and shorebirds such as the Wedge-tailed Shearwater (*Puffinus pacificus*), Fairy Tern (*Sterna nereis*) and Caspian Tern, with some species at the limit of their geographic range.*

The Grey-headed Honeyeater, a bird of rocky hillsides and gorges throughout much of arid Australia is one of four members of the Miner and Honeyeater family (Meliphagidae) found on Dolphin Island.

Being so close to the mainland, Dolphin supports a diverse avifauna, with 69 species recorded — second only to Enderby with 71 species. Dolphin is also home to the widest variety of mammals and reptiles found in the archipelago. But its proximity to the mainland also exacts a price. Both foxes and cats easily negotiate the narrow slice of Searipple Passage, frequently dry at low tide.[10] Over the past 50 years they have moved along Mermaid Sound and have been implicated in the decline of the Rothschild's Rock-wallaby on Dolphin Island.[11] They are an obvious threat to many other species.

Rich Minerals, Pearls and Desert Peas

Two runabouts sped past Flying Foam Passage's pearl farms, marked out on the waters below in row after row of neatly spaced black buoys. Gone are the miseries of last century, but pearling still extracts wealth from these waters. The boats zipped along, keen to reach some spot prized by the locals, and there settle for a pleasant afternoon's fishing. As industry grows, so does the region's population. With more people, the archipelago has assumed an ever greater importance as a place of recreation.[12]

To the north-east, Dolphin tapered into a breathless sea behind the 15 kilometre-long natural breakwater of Legendre Island. Most islands in the Dampier Archipelago are designated as nature reserves. Legendre is not in their number. It remains a "Ministerial Reserve for Industrial Development", earmarked for the possible future construction of a deep-water port.[13] If Legendre were to be transformed, road and rail access would be possible only across the Burrup Peninsula, Dolphin and other islands, and right through the Dampier Archipelago Nature Reserves.

The year 1999 will mark 300 years since William Dampier saw the archipelago that now bears his name. His words expressing a half-hearted desire to explore further in search of "some Sort of rich Mineral" now seem prescient. If the quest for "rich minerals" remains a priority, attended inevitably by roads and runabouts, ships, ports, pipelines and the potential for pollution, then so must we value equally, Sturt's Desert Pea flowering on an ocean beach.

1   Evidence suggests, if somewhat inconclusively, that the specimen of Sturt's Pea collected by Dampier may have come from the island, if not the very beach, on which we landed. Dampier named his landfall Rosemary Island but most likely went ashore on East Lewis or perhaps Malus Island. Dampier's description of *Roebuck*'s anchorage and the present day Rosemary Island don't correspond. The assignment of "Rosemary" to the island that now appears on charts with that name was the result of a cartographical error made by Louis de Freycinet, navigator on Baudin's *Le Géographe* Expedition. Phillip Parker King in 1818 deduced that Malus Island was Dampier's landfall.

2   W. Dampier, *A Voyage to New Holland*, London, 1729 (edn J. Spencer [ed.], Allan Sutton, Gloucester, 1981), p. 172. Dampier's specimens still survive and reside in the Sherardian Herbarium at the University of Oxford. Dampier's "Rosemary" was a species of *Olearia* related to the common coastal daisy *Olearia axillaris*.

3   *Ibid.*, p. 172.

4   *Ibid.*, p. 172.

5   Dampier is well known for his observations of the Aborigines of the Kimberley peninsula that would later bear his name. On his first visit to Australia he described them as the "miserablest people in the world". *A New Voyage Round the World*, vol. 1, 1697.

6   P. P. King, *Survey of the Intertropical and Western Coasts of Australia*, vol. 1, p. 37.

7   J. B. Gribble, *Dark Deeds in a Sunny Land*, Daily News, Perth, 1905 (repub. University of WA Press, 1987), pp. 47–8.

8   Enderby Island is slightly larger, with an area of 3,290 hectares.

9.   Summer temperatures rise to a February mean daily maximum of more than 36° Celsius.

10   The Burrup Peninsula is itself really an island, joined to the mainland by tidal mudflats and in recent times a causeway. Before the industrial facilities were built on the peninsula it was known as Dampier Island.

11   The Department of Conservation and Land Management (CALM) has been controlling foxes on Dolphin Island for the past ten years or so. This has led to a resurgence in the population of the Rothschild's Rock-wallaby.

12   The regional population (Shire of Roebourne) grew from 1,000 in 1965 to 18,000 in 1991. K. Morris, *Dampier Archipelago Nature Reserves Management Plan*, CALM, 1990, p. 16.

13   *Ibid.*, p. 21.

DAMPIER ARCHIPELAGO

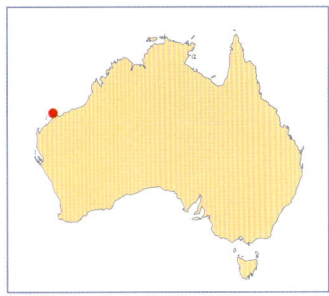

LOCATION

20° 20' S to 20° 45' S and 116° 25' E to 117° 05'E. 40 islands, islets and rocks within a 45 km radius of the port of Dampier on the Pilbara coast.

AREA

Enderby Island: 3,290 ha
Dolphin Island: 3,203 ha
Angel Island: 880 ha
Malus Island: 246 ha

CLIMATE

Semi-desert, tropical

STATUS

25 of the islands in the Dampier Archipelago are nature reserves. The remainder are Crown Land or mining leases. The Dampier Archipelago Nature Reserves are administered by the Department of Conservation and Land Management (CALM), Western Australia.

ACCESS

Access varies with zoning within the Dampier Archipelago Nature Reserves, eg on North Malus, Keast and all but the sand beaches on Hauy Island, no public access is allowed in order to protect wildlife populations. On most other islands, day use and beach camping permitted. Contact CALM for details.

FACILITIES

None

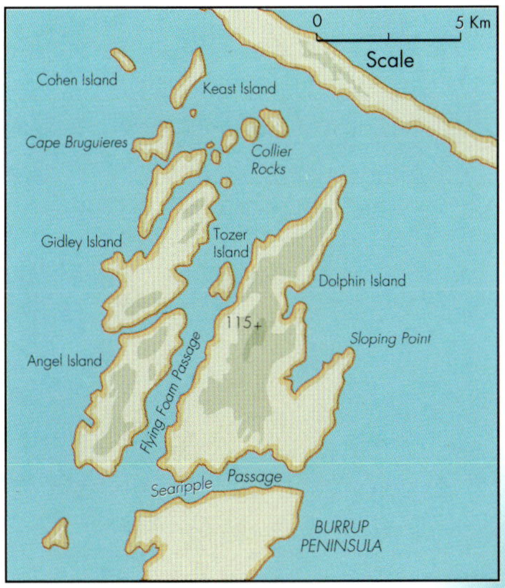

ISLANDS OF FLYING FOAM PASSAGE

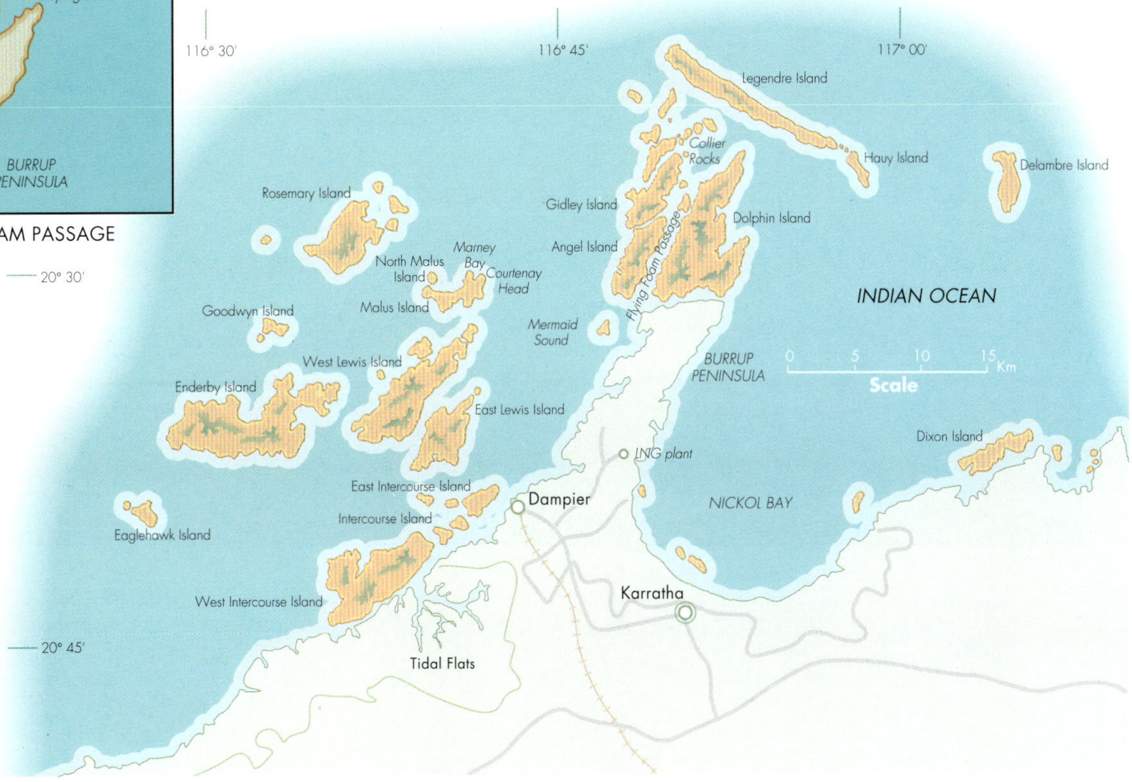

0   5 Km
Scale

Cohen Island
Keast Island
Cape Bruguieres
Collier Rocks
Gidley Island
Tozer Island
Dolphin Island
115
Sloping Point
Angel Island
Flying Foam Passage
Searipple Passage
BURRUP PENINSULA

20° 30'

Legendre Island
Collier Rocks
Hauy Island
Delambre Island
Rosemary Island
Gidley Island
Dolphin Island
North Malus Island
Marney Bay
Courtenay Head
Angel Island
Flying Foam Passage
Goodwyn Island
Malus Island
Mermaid Sound
INDIAN OCEAN
BURRUP PENINSULA
0   5   10   15 Km
Scale
West Lewis Island
Enderby Island
East Lewis Island
LNG plant
Dixon Island
NICKOL BAY
East Intercourse Island
Eaglehawk Island
Intercourse Island
Dampier
Karratha
West Intercourse Island
Tidal Flats

20° 45'

116° 30'   116° 45'   117° 00'

Shore Terrace (Christmas Island)
Oil on canvas 107 x 91cm

FATHOMS TO FORESTS

CHRISTMAS ISLAND

*Shunted from the deep ocean floor over
long intervals of geological time,
Christmas Island's verdant rainforest
terraces host some of the most
spectacular natural phenomena and
rarest creatures on earth.*

CHRISTMAS ISLAND

"... south by Java Head there is only Antarctica and Christmas Island ..." [1]

Tropical fish gathered in the intense aqua shadows of an undercut sea-cliff. Intermittently, the ocean surge swept them out into dappled light where they hung for an instant, resplendent in the flash of their gaudy array. Below their restlessness, a shallow rock-shelf studded with corals sloped away for a short distance before it tumbled into the ultramarine depths. Not far north from Christmas Island, the ocean floor bottoms out seven kilometres down in the Java Trench, one of the deepest ocean abysses on Earth.

Suspended above the rim of this darkening void, I snorkelled in the drift, lazily content to be pushed back and forth in the confused swell reflecting from the cliffs. I looked up to gather my bearings, only to find that I had floated some way out from the shore. At that distance, the waterline appeared topped by fretted grey bluffs and pinnacles running towards each horizon in a rough, unbroken band. In total contrast, a dense mantle of rainforest sprang from behind, climbing uninterrupted to the skyline.

From my shifting vantage point just off the north-west coast, there was nothing in any direction but the island or open ocean. The nearest landfall, the western end of Java flanking the Sunda Strait, was 360 kilometres to

Opposite: A large sea cave in the undercut cliffs near North West Point. Limestone cliffs such as these buttress the island in an almost unbroken band up to 45 metres tall.

Below: Long Indian Ocean swells assault jagged limestone pinnacles at the Blowholes on Christmas Island's south coast. Conveyed at six to seven cm a year on the Indo-Australian Continental Plate, the island is grinding towards eventual demise in the depths of the Java Trench, approximately 140 km to the north. At current rates this journey may take a further two to three million years.

the north. To the south, only Antarctica, half a world away.

Abrupt limestone walls like those arrayed ahead, comprise all but a fraction of Christmas Island's 73 kilometres of coastline. Despite rising on average no more than 10 or 15 metres out of the long Indian Ocean swells, these spray-soaked fortifications encircle the island, emphatically guarding its isolation.

A Virgin Deflowered

After the earliest known European sighting made in 1615 from the English East India Company ship *Thomas*, the unwelcoming coastline discouraged human activity for more than 350 years. A second recorded sighting occurred on 25 December 1643. Captain William Mynors aboard the *Royal Mary* named the island in honour of the day but did not land. In 1688, the *Cygnet*, en route to England with William Dampier aboard, called at Christmas Island after exploring the north-west coast of Australia. *Cygnet*'s crew made the first recorded landing: "*It was deep Water about the Island, and therefore no Anchoring: but we sent two Canoas ashore to search for fresh water ... and one found a fine small Brook near the S.W. point of the Island; but there the Sea fell on the Shore so high, that they could not get it off.*"[2]

In the 1820s, Englishman John Clunies-Ross settled 900 kilometres south-west of Christmas on a remote coral atoll known as the Cocos (Keeling) Islands. Here, Clunies-Ross and his family developed extensive coconut plantations with the sweat of Malay labourers. For ships travelling between Singapore, Batavia and Cocos, the virgin Christmas Island proved a convenient stopover. Clunies-Ross's men often took timber, seabirds and crabs from the island. Even loads of soil were shipped away to help establish the Cocos plantations.

Not until 1887 was the island systematically explored, when a party from the British survey vessels, HMS *Egeria* and HMS *Flying Fish* managed to reach the central plateau. Responses remained as unenthusiastic as Dampier's. However, rock samples were collected at the behest of the renowned naturalist Sir John Murray and taken to Britain on the *Flying Fish*. His analysis of the samples back in Scotland radically altered interest in Christmas Island: the rocks proved to be pure phosphate of lime — fertiliser for the agricultural hunger of the British Empire.

Within a year, the island had been annexed by Britain, with a settlement established at Flying Fish Cove by Andrew Clunies-Ross and a party of Cocos Malays. A lease was granted over its phosphate reserves in 1891, with Murray and Clunies-Ross's brother, George, as founders and joint investors in the Christmas Island Phosphate Company. Open-cut mining commenced in earnest in 1899 and phosphate was soon leaving for Britain and Australia by the shipload. Administratively, the island became part of the colony of Singapore.

In an era when the natural world was viewed almost exclusively as a resource ripe for exploitation, it was fortuitous that Murray had a scientific background. In 1897, he commissioned C. W. Andrews of the British Museum to conduct a thorough natural history survey. Andrews paid a return visit in 1908, documenting changes from the intervening years. From these surveys, a rare and clear picture of a pristine island was gleaned at the very time it suffered the initial impacts of human occupation.

Labourers were brought from Singapore, Java and Ambon to toil in the forest, clearing the tree cover and stripping phosphate rock by hand. The industry and the island's population grew slowly until the calamity of World War II. Invasion by the Japanese in 1942 led to the cessation of mining. Most Europeans were evacuated but the remaining islanders were stranded, forced to forage in the forest for food.

At war's end, the British-owned Christmas Island Phosphate Company was sold to the Australian and New Zealand governments, and in 1958 sovereignty itself passed to Australia.

Except for the period of Japanese occupation and two years in the late 1980s, mining has proceeded unabated, moving into the western arm of the island. However, in 1987 all forest clearing was stopped. Access is now permitted only to existing phosphate stockpiles and previously worked land. Despite 90 years of forest stripping prior to mining, about 75 per cent of the island's magnificent green cloak remains pristine. In 1980 a national park was declared over more than half of the island It has been twice enlarged since and further extensions are planned.

Up from the Deep

About 60 million years ago, a volcano began to grow from a fissure on the ocean floor, more than 2,500 kilometres south of Christmas Island's current position. Ten million years passed as the volcano accumulated, before it finally broke the ocean surface. But as this submarine mountain rose, the crustal plate on which it rode also inched north, pushing the foundations of the island to within 700 kilometres of its current location. By then, the volcano's upper slopes were bathed in light and warm tropical water. A coral atoll was born. Reef turned to limestone. On the growing undersea platform, future economic wealth was gradually secreted away in the form of calcium phosphate, laid down as marine deposits among the coral gardens of the embryonic island.[3]

In the most recent 10 million years, Christmas Island's distinctive stepped and terraced topography emerged as successive episodes of uplift pushed what were once coral reefs several hundred metres in the air. Fringing reefs continued to grow as new land appeared, stranded above the swell. Sea-cliffs became lofty heights, clad in rainforest. With the push towards the tropics, Christmas Island had undergone a mutation in the eternity of geological time from flat coral atoll to craggy and verdant oceanic island.

In its present location at 10°35' South, Christmas enjoys the typical climate of the wet–dry tropics, with two distinct seasons. From December to April the gusty north-west monsoon brings heavy rain and an occasional violent storm. With a reversion to steady south-easterly trade winds, a pronounced dry season sets in for the remainder of the year. Daily temperatures stay within a narrow, comfortable range of 20–30° Celsius throughout, with relative humidity remaining high due to the maritime influence.

Above: The calyxes of fallen mangrove flowers at Hosnie's Spring.

Opposite: Well inland and 30 metres above sea level, the mangroves Bruguiera gymnorrhiza *and* B. sexangula *grow as towering forest trees on a small basalt outcrop at Hosnie's Spring. The constantly watered half-hectare stand is thought to have persisted since the area was at sea level and may be up to 120,000 years old. Hosnie's Spring is recognised as a wetland of international significance.*

Above: Abbott's Booby chick. The entire breeding cycle takes about 16 months from nest building to fledging. The single, nest-bound and helpless chick is extremely vulnerable to being hurled to the ground in storms or cyclones. On average, five years are needed for an adult pair to raise a surviving independent juvenile.
Holger Rumpff

Right: A male Brown Booby, distinguished by the grey-blue base to his bill. There are an estimated 5,000 to 7,000 Brown Booby pairs nesting on the island.

Below: The perpetually damp ground at Hosnie's Spring and other freshwater soaks is ideal habitat for the Blue Crab, a widespread but vulnerable Indo-Pacific species. It is fully protected on Christmas Island.

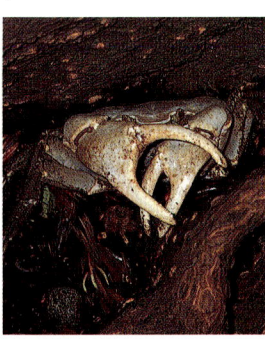

Boobies and Bosunbirds

Snorkelling not only gave me a marvellous opportunity to glimpse the inshore marine life, it also came with a dress-circle view of one of the world's great seabird congregations. Like so many isolated specks in the open ocean, Christmas provides a breeding station for great numbers of seabirds. Eight species nest on the island. Three of these are endemic: the Christmas Island or Andrews' Frigatebird (*Fregata andrewsi*), Abbott's Booby (*Sula abbotti*) and the magnificent Golden Bosunbird (*Phaethon lepturus fulvus*), a sub-species of the White-tailed Tropicbird.

Near the edge of the sea-cliff among a maze of weathered limestone, a downy-white chick waited expectantly for a parent returning from sea. The skies above were crowded with flight. Swooping close to the edge with exhilarating grace, a Brown Booby (*Sula leucogaster plotus*) suddenly swerved, extended its enormous webbed green feet and deftly flopped down beside its offspring. Nearby, more adults sat tight on their eggs. Widespread on the island, Brown Boobies nest in loose colonies on the open shore-terrace and along the rim of the lowest inland cliff.

Another adult landed. Reunited with its mate, the two immediately engaged in a ritual session of bill sparring, interspersed with a chorus of wheezing and honking — the confirmation of their bond. Breeding throughout the year, both adults share parental duties. They remain close to the island, so each

is readily able to alternate duties as they intersperse feeding with nest building, incubation and the care of young.

More than just a verdant blanket of green, the entire forest canopy behind the coastal cliffs was flecked with the flapping white forms of perched or nesting Red-footed Boobies (*Sula sula rubripes*), the most numerous Christmas seabird. The shoreline forest is home to some 12,000 pairs usually found roosting high on ambitious tree-top nests of twigs and sticks. Unlike the Brown, the Red-footed Booby is a true pelagic feeder, foraging at night far out to sea, searching for flying fish and squid.

In the breeze that had picked up from the north-east, great squadrons of frigatebirds soared overhead. Predominantly black, with sleek angular wings, deeply-forked tail and long hooked bill, theirs was a quite sinister manifestation. Hanging in mid-air, as though stalled, they waited to harry and harass other seabirds. With a dramatic change of speed two birds swooped. Their target was sent into a wild spin until it regurgitated its prey. The victim's meal was then pounced on by one of these aerial pirates while still in free fall. Despite these ruthless acts of piracy, most of the frigatebird's catch of squid and flying fish is snatched with its formidable bill, straight from the surface of the sea.

The endemic Andrews' Frigatebird is

Right: Christmas Island Hawk-owls (Ninox squamipila natalis) *roosting in the plateau forest among the endemic* Pandanus elatus. *Seven of the 13 landbirds found on the island are endemic species or subspecies. The insectivorous hawk-owl is the rarest, with a population of perhaps no more than 100 pairs.*

Previous page: Rainforest typical of the higher terraces. Although not as tall as the closed forests of the plateau, the terraces are floristically richer.

considered rare and endangered. Fewer than 2,000 pairs nest in well-defined colonies dispersed around the coastal treetops. In addition to the Andrews' Frigatebird, the Great Frigatebird (*Fregata minor minor*) also breeds on Christmas. Widespread throughout tropical seas, more than 3,000 Great Frigatebird pairs nest in the semi-deciduous trees of the shore-terrace, particularly around North West Point.

Frigatebird colonies are as extraordinary as their aerial displays. Widely spread wings quiver and flap as the birds perch on flimsy branches and the canopy resounds with screams and rasping siren-like calls. From far below, the occasional flash of brilliant crimson can be spied — not some showy tropical flower but the male frigatebird's balloon-like gular, or throat pouch. In eternal hope of a mate, males inflate these flamboyant elastic "shirt-fronts" whenever females fly close.

Seabirds in the Trees

Descending from the plateau forest to the shore-terrace behind West White Beach, my first impressions of Christmas Island's forests were shaped as much by the bizarre assortment of sounds emanating from within, as they were by the sumptuous green enclosure arching overhead.

Somewhere high up in the trees, the Christmas Island Imperial Pigeon (*Ducula whartoni*) let out its deep whoo, whoo, whoooo, birds calling back and forth in their persistent, advertising voice. But a loud and deep guttural croaking, also from above, was the oddest of all sounds. Luckily, I was able to see past a break in the canopy through to the crown of an emergent tree. There the author of that strange call straddled a substantial woven nest built atop a flimsy branch: an Abbott's Booby, the largest and rarest of the nine seabirds in the family Sulidae.[4]

Nesting 25–40 metres up in the canopy of the central plateau, these huge pelagic birds appear totally incongruous perched in the tangle of the treetops. In a biological trade-off to gain maximum range for minimum effort, these avian "jumbo jets" are equipped with a huge wingspan. Their great bulk and long narrow wings are perfect for gliding effortlessly above the ocean wave, but when it comes to landing and take-off, they are a decided liability. On take-off, an adult may drop as many as 10 metres before achieving flight, and landing can only be attempted into the wind. The flimsy lateral branches just below the crown of emergent trees provide the only suitable roosting place for the Abbott's Booby.

Due to loss of habitat, they now only nest on Christmas Island, where there are perhaps just 2,000–3,000 breeding pairs. A de facto endemic, their former breeding distribution remains an enigma. Decimated by hunting and habitat destruction, they certainly once bred on a number of tropical Indian Ocean islands and recent evidence suggests that their range may have extended across the Pacific.

Their continued survival on Christmas itself has been directly threatened by loss of habitat. Concentrated on the western end of the island, the nest sites of the Abbott's Booby coincide with phosphate deposits. The birds remained virtually undisturbed until the 1970s but by 1974 mechanical clearing had destroyed a quarter of their forest home. An attendant population decline of about 15 percent, led to the obvious conclusion that further destruction of plateau forest would threaten extinction.

In the shelter of the forest, I had not been aware of the stiff sea-breeze blowing over the island. The tree-top aerie of the Abbott's Booby appeared even more precarious, as the whole crown suddenly swayed and gyrated in a strong gust. Clearing on the plateau has not only directly deprived the birds of nest trees, it has made downwind areas of surviving forest more prone to wind turbulence. Living at such a great height, already vulnerable chicks are in great danger of being hurled from the nest, condemned to starve on the forest floor or fall victim to marauding crabs.

In 1989, an extensive revegetation program on mined sites was begun.[5] Aimed at returning all cleared areas to near natural forest cover, nearly 100,000 trees had been planted by 1994. The urgent priority has been rehabilitation adjacent to Abbott's Booby nest sites in an attempt to reduce downwind turbulence. Fast-growing plants, such as the exotic Japanese Cherry (*Muntingia calabura*) and the native *Macaranga tanarius*, are preferred as pioneer plants.[6] They form a quick protective canopy for slowly developing indigenous species which in turn progressively shade out the dwindling pioneers. Stabilisation comes first. The natural succession of the island's forest may take more than 100 years.

Christmas Rats

While the fate of the Abbott's Booby is not entirely sealed, that of most of Christmas Island's mammals unfortunately is. In an environment where crabs dominate the terrestrial niches of the rainforest to the total exclusion of ground-nesting birds, the existence of mammals in the same realm seems improbable. However, before the arrival of the *Flying Fish*, the forest floor not only swarmed with crabs but with two species of endemic nocturnal rat as well: Maclear's Rat (*Rattus macleari*) and the Christmas Island Burrowing Rat (*Rattus nativitatis*).

C. W. Andrews described Maclear's Rat from his 1897 visit: "In every part I visited it occurred in swarms. During the day nothing is to be seen of it, but soon after sunset numbers may be seen running in all directions, and the whole forest is filled with its peculiar querulous squeaking ..."[7] Yet when Andrews returned in 1908, he was unable to find a single specimen of either species. That hitchhiker of the sea, the Black Rat (*Rattus rattus*) reached the island with the first settlers and probably brought along a cocktail of microbes that the native rats were unable to resist.

Luckily, two species of bat, the fruit-eating Christmas Island Flying Fox (*Pteropus melanotus natalis*) and the tiny insectivorous Murray's Pipistrelle (*Pipistrellus murrayi*) have survived.

Above: The Robber or Coconut Crab is well adapted to a terrestrial life, readily climbing the loftiest Arenga Palm in search of fruiting delicacies or scrambling several metres up the nearest tree when disturbed. It is the world's largest land crab—the length of an adult's body (excluding legs) often exceeds 200 mm.

Below: The Red Crab plays an important role in the nutrient cycle of the island's forests, accelerating the breakdown of nutrients otherwise held in forest-floor debris. Their feeding restricts seedling survival, while their burrows aerate the soil. Although considered an endemic, very similar specimens are found on North Keeling in the Cocos (Keeling) Islands.

The flying fox is active in daylight and can be seen in large numbers, sweeping and soaring on the updrafts above the inland cliffs. The fifth of the endemic mammals, the insectivorous Christmas Island Shrew (*Crocidura attenuata trichura*) remains a mystery. It has not been sighted since 1985 and may now also be extinct.

Magnificent Simplicity

Wherever I walked, the forest floor had a well-tended look about it, almost like a corner of some city park. Progress was easy. It wasn't the idea of the steamy tropical jungle that I had expected. The great diversity usually associated with Asian rainforests is not found here; just 135 species, 24 of which are trees, make up the entire suite of indigenous plants. But in common with an oft-observed insular phenomenon, the island's trees grow larger, and seemingly with more vigour, than their continental relatives.

Climax forest grows on the deep soils of the central plateau. Here great spreading, buttressed trunks of the Kelat (*Syzygium operculatum*) and other emergent trees pierce the canopy which itself rises to a near entire green vault, 40 metres off the ground. Stately 50 metre-tall Tahitian Chestnuts (*Inocarpus fagifer*) are prominent, lining water courses and towering over wetter sites.

Stepping down the terraces, forest height varies with soil depth. On shallow soils, exposed limestone pinnacles erupt from the forest floor like avant-garde sculptures. On these intermediate levels sinuous Banyans (*Ficus microcarpa*) are a feature. The forest is lower, but floristically more complex than that on the plateau.

Exposed to withering, salt-laden winds, the slopes of the shore-terrace support a diverse deciduous forest. Common species include: Ketapang (*Terminalia catappa*), Propeller Tree (*Gyrocarpus americanus*), the Coral Tree (*Erythrina variegata*), Cottonwood or Beach Hibiscus (*Hibiscus tillaceus*), and pure stands of *Pisonia grandis*, that great tree of tropical seabird islands worldwide. Many species are pan-tropical and shared with the Indonesian and Malay archipelagos, or with northern Australia.

Terrace trees dwindle as the inhospitable cliff-edge is neared. Here, only hardy shrubs and herbaceous plants persist. Shiny, apple-green leafed *Scaevola sericea*, common to Indo-Pacific shorelines, forms compact thickets among spray-saturated pinnacles, and the rattle of stiff, strap-like leaves of the endemic *Pandanus christmatensis*.

Conspicuous in the forest are several of Christmas Island's 16 endemic plants, among them the graceful palm *Arenga listeri* and the second of the two pandanus species, *Pandanus elatus*. Less obvious are the Christmas orchids. Four of the 11 are endemic but most are canopy-dwelling epiphytes and therefore hard to spot. An exception are the delicate, crimson-splotched white blooms of *Brachypeza archytas*, frequently found low on the boles of terrace giants.

Crustacea

The well-tended appearance of the forests is not due to some phantom team of gardeners but is entirely the work of Christmas Island's most famous residents: Red Crabs (*Gecarcoidea natalis*). These ultimate composters cycle through their bodies anything remotely edible that falls to ground. Leaf-litter, fruits, flowers, seeds, dead seabirds and the carcasses of less fortunate cohorts, all are assiduously consumed by this army of crustaceans.

The nutrient cycle of the forest is inextricably bound to the activities of a stupendous adult population — an estimated 120 million crabs. In the crabs' preferred habitat (the tall forests growing on deep soils), a biomass of one tonne, or 13,800 crabs per hectare, crowd the densely shaded ground.

With the coming of the first wet season rains (usually in late November), the crabs begin their astounding migration from forest to sea. Moving in close synchronisation with the lunar cycle, the entire adult population descends in waves to mate on the shore-terrace. Here, males excavate burrows which they defend fiercely against occupation from their brothers, before the female is courted and lured below.

Copulation usually occurs in subterranean privacy. Males then dip in the sea, replenishing their body fluids, after which most return inland.

Meanwhile, the female remains below ground for nearly two weeks while her clutch of up to 100,000 eggs develops. On the last quarter of the moon, a seething congregation of female crabs descends to the sea and releases great clouds of spawn into the night's high tide. The eggs hatch immediately into incalculable numbers of crab larvae known as megalopae. For hundreds of millions of megalopae this is the start and finish of a short life. After about 25 days the tiny survivors return to the shore, swarming as a thick red scum over the rocks. Once at the tideline, the megalopae metamorphose into baby crabs and then begin the journey inland. In years when mortality is relatively low, a near seamless carpet of little crabs moves across the island as far as the plateau, over 200 metres above and several kilometres inland from their birthplace. Settling in the first available safe hideaway, they lead a secretive life for the following three years, before they themselves join the main community and the annual migration back to the sea.

In sheer numbers, the Red Crab has no rival. However, it is only one of 20 species of land crab found on Christmas Island. Crabs occupy every niche from the intertidal zone to the plateau heights. Two species, the Blue Crab (*Cardisoma hirtipes*) and as its name implies, the Freshwater Crab, (*Ptychognathus pusillus*) are adapted to life around soaks and streams. The Christmas land crab community is probably the largest and most diverse in the world.

Sharing all parts of the Red Crab's domain, the enormous Robber or Coconut Crab (*Birgus latro*) is a common sight, creeping around the forest floor. Robbers are often seen climbing trees — the fruits of the Arenga Palm are an eagerly devoured favourite. Of worldwide tropical distribution, the Robber Crab has long been considered very good eating. William Dampier attested to its culinary worth in 1688, when members of his crew brought a quantity back to the *Cygnet*:

"They got also a sort of Land-Animal, somewhat resembling a large Craw-fish ... These creatures lived in holes in the dry sandy Ground, like Rabbits. Sir Francis Drake in his Voyage around the World makes

Dolly Beach, one of the few sand beaches on Christmas Island, is used by small numbers of nesting Green and Hawksbill Turtles.

mention of such ... They were very good sweat Meat, and so large that two of them were more than a Man could eat; being almost as thick as ones Leg..."[8]

As with the Abbott's Booby it is extremely fortunate that the Robber Crab survives on Christmas Island. Threatened elsewhere this lumbering giant is protected within the bounds of the Christmas Island National Park, along with all of the island's indigenous fauna and flora. The interaction of humans with the Christmas environment is no longer the destructive relationship of years past.

Changing the Guard

The island today is home to a multi-racial population of about 2,500 people, predominantly of Chinese descent, with smaller groups of Malays and Europeans. Although the phosphate industry continues to contribute to the island's wealth, mining will soon come to an end. To ensure an ongoing life for the islanders, economic diversification is now being actively pursued, with tourism seen as the best future prospect. A resort hotel and casino opened in 1993 on previously cleared cliff-top land near Ethel Beach is the island's first significant development project outside the phosphate industry.

Some may feel disquiet at the thought of high rollers and denizens of the piano-bar out there among the frigatebirds and boobies. However, one percent of the casino's gross profit each year is directed to the benefit of the Christmas Island community and roulette wheels must be infinitely preferable to bulldozers and ore trucks. The challenge will be to adequately safeguard the island's natural splendours as both the population and visitor pressure grows.[9]

It is within the park, as a showcase of the wild Christmas Island, that the future really lies. Here, in dazzling display, the patterns of island ecology, colonisation and evolutionary change are on show for those with the will to see.

The flight back to Perth and that "other" Australia was delayed. The extra hour gave me a chance to quickly revisit a favourite spot. Margaret Knoll is a kilometre from the airport, and a short climb from the end of the track affords panoramic views along the east coast. The cliff-edge below the lookout is a well-used nesting site for Brown Boobies. Just as it was time for me to leave, a booby flew in, exchanged pleasantries with its mate and changed guard over the pair's single egg. A common enough domestic scene, but it left me with a sense of great privilege, that this island offers such ready and almost casual glimpses of nature untrammelled.

More like primeval Java than anywhere on the Australian continent, the island's astonishing natural phenomena have survived almost intact despite a century of disturbance. Christmas Island holds a special place in an Indian Ocean so bereft of land — yet on the edge of South East Asia's galaxy of islands, its importance is no less diminished. As the region's remaining pristine environments are inexorably crushed under the weight of humanity, the value of the gifts of Christmas lie even further beyond estimates of importance.

1   Quoted in M. Neale, *We Were the Christmas Islanders*, Bruce Neale, Canberra, 1988.

2   W. Dampier, *A New Voyage Round the World*, London, 1697, (edn Dover, New York, 1968), p. 159.

3   The original theory of the origin of Christmas Island's phosphate reserves proposed that they were deposited as guano, the droppings of countless generations of seabirds, when the island was a dry, low-lying coral atoll. While this source may have had a contributing effect, it is now discounted as the principal source, as phosphate deposits have been found in continental shelf deposits around the world.

4   The family Sulidae are the six Boobies of the genus *Sula* and the three Gannets of the genus *Morus*. Four of the six Boobies occur in tropical Australasian waters.

5   The Christmas Island Rainforest Rehabilitation Program (CIRRP) is administered by the Australian Nature Conservation Agency (ANCA) and funded by 36.6% of the royalty paid by the mining company.

6   Japanese Cherry is actually a native of the West Indies and had previously been introduced to the island.

7   Quoted in T. Flannery, 'Rats of Christmas Past', *Australian Natural History*, vol. 23 (5).

8   Dampier, *op. cit.*, p. 159.

9   The economic development plan sees a population of 16,500 by 2015. See *Expressions of Interest, Christmas Island Tourism and Associated Developments*, Wheelan Consultants, Perth, 1995.

CHRISTMAS ISLAND

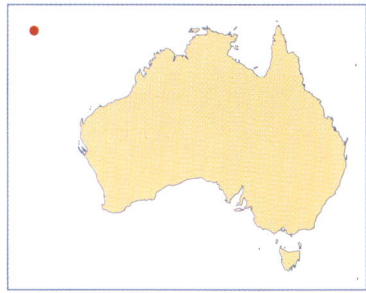

LOCATION
10° 35' S, 105° 40' E. Roughly equidistant, 2,600 km west of Darwin and north-west of Perth. 360 km south of Java and 1,400 km from the closest mainland town at Port Hedland.

AREA
13,500 ha

CLIMATE
Tropical, monsoonal

STATUS
Australian External Territory. 63% is national park (approximately 8,500 ha). 42 km of the 80 km coastline is marine park, extending 50 m seawards from the low water mark. National park extensions are planned. Administered by the Commonwealth Department of the Environment, Sport and Territories. Christmas Island National Park is managed by the Australian Nature Conservation Agency (ANCA).

ACCESS
Regular commercial and charter air services from Perth and Jakarta. Road, water and foot access to Christmas Island National Park. Bush camping is allowed with a permit—refer to ANCA.

FACILITIES
Christmas Island Resort and several lodges. No accommodation within Christmas Island National Park. Full town amenities—car hire, restaurants etc.

105° 35'

105° 40'

10° 25'S

North-East Point

settlement

FLYING FISH COVE

North-West Point

airport

Margaret Beaches

Casino Resort

West White Beach

Rhoda Beaches

Ethel Beach

Lily Beach

Hosnie's Spring

Margaret Knoll

Steep Point

The Dales

Murray Hill (361)

Winifred Beach

10° 35'

Greta Beach

Egeria Point

Blowholes

Dolly Beach

0   2   4 Km

SMITHSONS BIGHT

INDIAN OCEAN

Scale

Dorothy Beach

South Point

Christmas Is. National Park

Proposed park extensions

Sandstone Citadel (North Island, Sir Edward Pellew Group)
Oil on canvas 107 x 91cm

TROPICAL ARCHIPELAGOS

BONAPARTE ARCHIPELAGO

WESSEL ISLANDS

SIR EDWARD PELLEW GROUP

ISLANDS OF CAPE YORK AND THE TORRES STRAIT

Fortified by ancient sandstones, Australia's northern extremities have witnessed many migrations: from the sea's episodic advances and retreats, to the earliest arrivals of humankind on the continent. It was also along such shores that Macassans and Europeans first confronted an unknown southern land. To this day these remote, enigmatic islands remain one of Australia's great natural and cultural threshholds.

BIGGE ISLAND – BONAPARTE ARCHIPELAGO

Bay Watch

Boomerang Bay takes a large bite out of the west coast of Bigge Island. Bordered by jagged rocks and sickle-shaped beaches, the bay forms a natural opening into the heart of the island. Creeks draining from the interior converge here, forming an estuary verged by mangroves and mudflats. While our yacht tugged at its anchor in the bay, I was ferried ashore by dinghy. The coffee-coloured waters of the inlet streamed out to sea, drawn by the tide. There was an air of foreboding about this murky expanse and what might lie waiting among the mangroves.

It had taken us two and a half weeks of sailing from Wyndham to reach Boomerang Bay. The Kimberley coast is one of longest undefiled shores left on Earth. It winds for more than 3,000 kilometres, with fretted peninsulas and huge gulfs. Bigge Island is stationed roughly half-way along the coast, the second largest of the 1,000 or so islands distributed between Broome and Cambridge Gulf. These archipelagos collectively form the greatest mass of islands in Australia, and most are little-known and rarely, if ever, visited. We had woven our way cautiously through this maze, contending with huge tides, bullying easterlies and assorted reefs and rocks, some still to find a place on the charts.

Puttering our way into the estuary at Boomerang Bay, there were no obvious signs of life: no jaunty seabirds or darting fish visible in the turbid shallows nor any eyes emergent from the rippling surface of the bay. But what can be readily seen is rarely conclusive in such locales. When the dinghy stuck fast on the mud, I decided to continue on foot to the shore, across 200 metres of ankle-deep water and thick mud. I lurched in the direction of dry land, the mud clinging to my shins like warm fudge. The sight of two small sharks feasting on fish trapped in the shallows ahead, somehow, seemed less troubling than the possibility of other unseen eyes watching and waiting for me to stumble.

When I finally reached the shore I went to a rockpool to wash the mud from my legs. As I did so, I noticed a long groove running down the beach, across the mud and into the water. Either side of the groove were tracks left by four widely spaced reptilian feet—the tracks of the world's largest living reptile species, the Estuarine Crocodile (*Crocodylus porosus*).

The sense of isolation fostered by islands is nowhere more acutely felt than in the Kimberley. Here, even the adjoining mainland is itself abundantly remote from centres of 'civilisation'. As I padded along the shore, at the beginning of several days alone on Bigge Island, the solitude was all-embracing. Climbing inland past a bank of red earth embedded with layer upon layer of shells—a huge midden left by the Wunambal people who once inhabited the island—further heightened the disjunction with the external world and familiar registers of time and place.

The inundation of Kimberley river valleys at the end of the last ice age some 10,000 years ago transformed crests of high ground into coastal outliers, such as these in the remote enclave of Prince Frederick Harbour.
Quentin Chester

With its prolific displays of pink flowers, the Turkey Bush is an eye-catching presence in the stone country of Bigge Island.

I scrambled along a scrubby ridge, anxious to find water and a place for the night. The daylight seemed to be draining away as fast as the tide. As I pitched my tent on a narrow ledge, the bush reverberated with the cries of birds settling for the night. As I tried to sleep a huge moon rose over the island, illuminating the estuary.

Rain Supreme

Earlier that day, during a brief visit to a beach at the northern end of the island, I had witnessed the haunting expressions of the Kaiara, spirit beings borne across the sea. Caves abutting the beach are emblazoned with ochre paintings of figures in boats and huge haloed heads. One large image stuck in my mind, its wan features framed by radial bands of colour and dominated by black, pitiless eyes.

The Kaiara strongly resemble the Wandjina figures, the dominant rock art tradition of the Kimberley region. And like the Wandjinas, the Kaiara are all-powerful embodiments of forces that are both creative and destructive. Confined to the coast, they are thought to control the weather and, if angered, may retaliate with a cyclone or 'whirlwind'. When archaeologist Ian Crawford visited this site in 1963, a campfire accidentally spread close to paintings. Some of the Aboriginal men in the party believed the eyes of the Kaiara had been scorched and that retribution was sure to follow. Strong winds and rainclouds duly appeared. Crawford noted, "We left immediately."[1]

On my visit, in mid-June, working against the possibility of a Kaiara-inspired whirlwind was the fact that the dry season was well under way. Storms and rain were highly unlikely for many months. But come December or January this coast is assailed by violent thunderstorms. Virtually all of the region's annual quota of rainfall (around 1,000 millimetres) comes from these intense monsoonal deluges. Rainfall is generally higher on the elevated country immediately inland from the coast. But the weather is rarely predictable and the islands bear the brunt of cyclones that develop over the Timor Sea and sweep towards the coast, bringing brutal winds gusting in excess of 200 kilometres per hour and several days of unrelenting rain. Such is the fury of the Kaiara.

On The Plateau

The next morning I was relieved to find sweet-tasting water trickling between terrace pools, a legacy of Kaiara activity many weeks earlier. Lavender-flanked Wrens (*Malurus dulcis*) bathed nearby and a Pheasant Coucal (*Centropus phasianinus*) lumbered into the air from its perch, its long tail dragging awkwardly behind.

Two large watercourses converged at the head of the inlet. Mangroves (*Avicennia* spp. and *Rhizophora* spp.) dominated the tidal flats of grey mud and gravel. Above the range of the tide the rocky creekbeds wound inland forming verdant avenues of Cadjeputs (*Melaleuca leucadendra*) and Screw Pine (*Pandanus spiralis*). As I picked my way across the boulders, Red-winged Parrots (*Aprosmictus erythropterus*) screeched and chattered in the tree-tops. Flood debris deposited by wet season torrents hung from high branches. Climbing up the steep creek banks I emerged onto a spacious plateau, extending across the heart of the island.

Away from moist creeks and wooded vales, the vegetation on Bigge Island assumes the characteristic guise of Kimberley stone country: a sparse smattering of shrubs and small trees including *Acacia, Hibiscus,* and *Ficus* species. Scattered among the larger shrubs are large clumps of Hummock Grass (*Plectrachne* sp.). Navigating across the plateau one's eye fixes on such distinctive plants as the Turkey Bush (*Calytrix exstipulata*), with its showy displays of pink flowers and the wispy foliage of the Fern-leaved Grevillea (*Grevillea pteridifolia*) swaying in the sea breeze.

Even more arresting is the spectacle of isolated sandstone outcrops and mesa-like blocks dotted across the plateau. Such structures are the hallmark of the remote north-west. Weathered by the elements, they are often undercut on all sides, with recessed

overhangs and small caverns. I moved among these stone islands, taking refuge from the broiling heat and immersing myself in the shadows of the terrain and its past.

In one large shelter I found walls daubed with paintings, not of the Kaiara, but of elongated serpents, wallabies and cryptic winged creatures. I crawled inside and lay on a bench of cool, polished stone, gazing at images on the ceiling. Wasps hovered against the rock and from the slender cavities at the back of the overhang came the sound of bats rustling. The floor of the shelter was a dark silt, as fine as talc, inlaid with shell fragments, chips of bone and fire-charred pebbles—all lying undisturbed. It was as if the occupants of the shelter had left only days earlier.

Rocks Of Ages

The hewn stone formations of Bigge Island reflect the legacy of a geological history with few equals. The most common rock-type exposed on the island is King Leopold Sandstone, which together with Warton Sandstone and the Carson Volcanics, dominate

Unknown to science until 1978, the Monjon—the smallest of all rock-wallabies—is a ubiquitous though elusive creature of Bigge Island's sandstone outcrops. Jiri Lochman/Lochman Transparencies

Eroded sandstone cliffs in the Buccaneer Archipelago betray the long history of weathering that shapes the ancient rocks of the wet-dry tropics.

Above: The all-powerful Kaiara, the Wandjina figures of the coast, are a haunting presence in the caves and rock shelters of Bigge Island.
Andrew Burbidge

Opposite: Evening light on the sandstone cliffs of West Governor Island, at the entrance to Napier Broome Bay, on the Kimberley's far northern coast.
Quentin Chester

landscape along major joints and fractures. By contrast, the Carson Volcanics give rise to an undulating terrain of high, rounded hills and a dense cover of vegetation.

To sail the Kimberley coast is to be constantly confronted by these divergent landforms. Passing among the islands it was common, for example, to have a rocky sandstone island to starboard, while off the port beam, rose a tall volcanic isle. The former are fortified by ragged orange and buff-coloured cliffs and support only meagre vegetation, whereas the latter feature wooded slopes flanked by tumbling screes of dark rock and gullies of vine thicket. In the Bigge Group, the Maret and Montalivet islands to the north-west are volcanic in origin. Some 20 kilometres closer to shore, the Wollaston, Katers and Prudhoe islands are formed from King Leopold Sandstone.

The present-day configuration of the Kimberley coastline and islands developed 8,000–10,000 years ago. Rising sea levels inundated ancient valleys incised in the landmass by major river systems, to form a deeply indented shoreline of gulfs and harbours. Meanwhile, areas of elevated terrain became isolated as islands. In the process whole communities of plants and animals were cut off from the mainland. These include significant pockets of rainforest or vine thicket, such as those on South-west Osborn Island in Admiralty Gulf, and Augustus Island, off Port George IV.

At 17,952 hectares, Augustus is the largest island on the Kimberley coast and boasts a significant mammal fauna that includes the Little Rock-wallaby or Narbalek (*Petrogale concinnus*), Sugar Glider (*Petaurus breviceps*) and the endangered Golden Bandicoot (*Isoodon auratus*). The conservation value of such relic populations is inestimable, divorced as they are from the depradations of human activity and introduced species.[2]

No less significant are the outlying coral and sand cays such as the Lacepede Islands, Adele Island, Ashmore and Cartier Islands, Browse Island and the Rowley Shoals. As well as being rich reef ecosystems, these islands are important nesting sites for various seabirds, such as the Lesser Frigate Bird (*Fregata ariel*) and turtle species, including Green Turtles

the geomorphology of the entire central Kimberley Plateau. The sediments making up the sandstone were laid down in shallow seas around 1,800 million years ago. Volcanic activity around this time also saw the extrusion of flood basalts and the intrusion of vast underlying sills of dolerite.

Since its formation, the flat-lying strata of the Kimberley Basin have remained essentially stable, with only localised areas of uplift. As a consequence the landforms evident on the islands and adjacent coastline reflect patterns of sustained weathering and erosion that have unfolded over many millions of years. On sandstone surfaces this has resulted in a highly dissected terrain of cliffs and terraces. Gorges and crevasse-like ravines are incised across the

The islands of the north-west assume a multitude of guises, from low-lying fragments to lofty plateau remnants such as Steep Island, near Raft Point. Quentin Chester

(*Chelonia mydas*).

Moreover, the full extent of endemic and endangered flora and fauna on the islands of the Kimberley is not yet known. A great many outliers and associated marine habitats are still to be surveyed. This 'tropical arid' region remains a rich frontier whose intrigues have only been partially fathomed by science.

Life Under a Bigge Fig

After a long day on the island's baking terraces I retreated back to a freshwater creek and pitched my tent high on an outcrop overlooking the estuary. This stack of blocks and ledges was shaded by a monstrous Rock Fig (*Ficus platypoda*), its pale grey limbs coiled with vines. This solitary tree, seemingly sprouting from the rock itself, was a hub of life. Battalions

of Green Tree Ants (*Oecophylla smaragdina*) marched up the root buttresses to cocoon-like nests hanging festively on the outer branches. On the opposite side of the tree a pair of Red-winged Parrots feasted on figs, while overhead a Great Bowerbird (*Chlamydera nuchalis*) leered down and serenaded me with its flatulent babble.

At dusk the outcrop became a hive of activity as small rock-wallabies emerged from their hideouts and scurried among the ledges and the surrounding undergrowth. These were Monjons or Warabi (*Petrogale burbidgei*), a species endemic to the extreme north-west Kimberley and unknown to science until 1978. Mostly nocturnal and incurably shy, this is one of least understood Macropods, with large round eyes as striking as those of the Kaiara paintings.

The sky above the inlet was flecked with bats,

strafing the mangroves for insects. A large group congregated in the crown of the fig, where they gnawed on fruit. I lay quietly, staring up at the tree's great spread of limbs and listening to the soft clap of wings. When I woke the bats were gone. A lone figure was stalking along a horizontal limb, an Illungalya or Scaly-tailed Possum (*Wyulda squamicaudata*), yet another enigmatic forager in the fig tree.

The New Kaiara

All of these elusive creatures, and many more, would have been familiar to the Aborigines who once occupied this island. Few details are known of tribal life on Bigge. Given the available land area, semi-permanent freshwater supply and abundant food sources, the island may well have supported long-term habitation. The rocky shores were a ready source of shellfish, while hunting expeditions in the bays and inlets would have yielded turtles, Dugongs and a variety of fish. From the land they would have harvested a variety of root crops and assorted reptiles, rodents and marsupials. The flighty responses of larger game like the Monjons and Illungalyas suggests a history of being hunted, both by Aborigines and Dingoes (*Canis familiaris*), which were brought to the Bigge with the tribes, and continue to roam the island.

The only lingering traces on the island of this culture are the middens, rock shelters and the Kaiara. Dingoes feature in these paintings, as do small figures, which according to traditional accounts are seen eating lily-roots and carrying berries. But to Western eyes they are the unmistakable images of boats and pipe-smoking sailors. These paintings seemingly represent an early attempt to merge the arrival of Europeans with the transcendent mythology of the Kaiara.

It has been suggested that such paintings refer to the first documented voyage in these waters made by Abel Janszoon Tasman in 1644. He commanded a fleet of three ships and charted a large section of the coast, including islands in the vicinity of Bigge.[3] Notwithstanding this possibility, it is likely that, over the centuries, the Kimberley coast was the

scene of many unrecorded shipwrecks and chance encounters with vessels from distant lands.

The first significant interaction between the Aboriginal islanders and the outside world was the result of visits by Macassan sailors. Arriving in search of trochus shell and trepang, they set up camps along the Kimberley coast.[4] There were repeated and sometimes violent conflicts with the Aboriginal groups. The Aborigines did, however, obtain Macassan dugout canoes, which greatly enhanced their ability to navigate among the islands.

Following Tasman there were several brief exploratory visits to the Kimberley coast. William Dampier made his famed Australian landfall of 1688 in the Buccaneer Archipelago to the south. Then, French expeditions in 1801 and 1803, both led by Nicholas Baudin, skirted the islands and attached names to numerous features. But not until Phillip Parker King began his detailed survey in 1818 was the true complexity of this coastline revealed. King undertook four separate voyages to complete

HM Cutter Mermaid, *the vessel captained by Phillip Parker King during his remarkable voyages of exploration along the treacherous Kimberley coast. Drawing by P.P. King, courtesy Mitchell Library, State Library of N.S.W.*

Scale 20 feet to an inch.

his majesty's Cutter Mermaid - L.t P. P. King 1817

his charting of these convoluted shores. He worked from east to west and Montague Sound was not reached until his third voyage aboard HMS *Mermaid* in 1820. He sailed through Scott Strait which separates Bigge Island from the mainland, and then proceeded south to explore the majestic Prince Frederick Harbour.[5]

Although King's discoveries aroused considerable interest in the region, it was the area around the Prince Regent River and parts south that attracted the most attention. In 1837 John Lort Stokes aboard HMS *Beagle* examined King Sound, the Buccaneer Archipelago and Collier Bay, while George Grey made his hapless inland foray in the Glenelg River Region. In the meantime, Bigge Island and the other outliers of Montague Sound and Admiralty Gulf were left in relative obscurity.

Given this neglect, it is almost certain that the Wunambal people continued to live undisturbed on the coast and islands well into this century. Inevitably, however, the dislocation of Aboriginal society across the Kimberley reached the remote north-west. Most of the remaining coastal inhabitants ended up in mission settlements like those at Kalumburu and Kunmunya. For nearly 50 years islands like Bigge have been left abandoned, their secrets guarded by the eyes of the Kaiara.

The Bigge Picture

During my final two days on the island I made more trips into the hinterland from my camp. Each sortie served to deepen the Bigge conundrums. I visited many rock shelters, some tucked beneath immense tilted slabs. From their summits the views across the bay suggested a landscape untouched since creation. But then scrambling into the cool recesses of the overhangs I was met by the glaring visages of the Kaiara.

The art sites on the southern extremities of the island gave the impression of having been abandoned since the last of the Aboriginal inhabitants left these shores. It was an exquisite sensation to be alone in such enclaves, surrounded by sublime artistic expression that had its roots in the deep past and the forces that shaped the lives of the island's inhabitants. At the same time my presence had a hint of trespass.

I walked from the stone country back to Boomerang Bay. The sky was clear and a soft breeze brushed through the beach spinifex. Over a headland lay the dinghy hauled up on the beach and nearby the yacht bobbing at anchor in the bay. I felt a sense of release from the pressures of the island. But at that moment I did not know that, as we sailed south, clouds were going to blanket the sky and whirlwinds would raise huge waves in our path.

---

1   I. M. Crawford, *The Art of the Wandjina*, Oxford University Press, Melbourne, 1968, pp. 69–80.

2   A. Burbidge, N.L. McKenzie and K.F. Kenneally, *Nature Conservation Reserves in the Kimberley*, Department of Conservation and Land Management, Perth, 1991.

3   Crawford *op. cit.* pp. 77–78.

4   See note 7 page 82.

5   Bigge Island was named by King after John Thomas Bigge who had earlier conducted a commission of inquiry into the colony of New South Wales.

BIGGE ISLAND – BONAPARTE ARCHIPELAGO

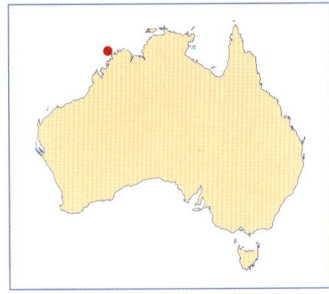

LOCATION
14° 30'S 125°10'E.
Approximately 330 km west-north-west of Wyndham.

AREA
17, 850 hectares

CLIMATE
Tropical, monsoonal

STATUS
Vacant Crown Land. Recommended by the Department of Conservation and Land Management for declaration as a Class A reserve for the Conservation of Flora and Fauna.

ACCESS
Private vessel or charter boats via Broome, Derby or Wyndham. Yacht anchorages in Boomerang Bay.

FACILITIES
None

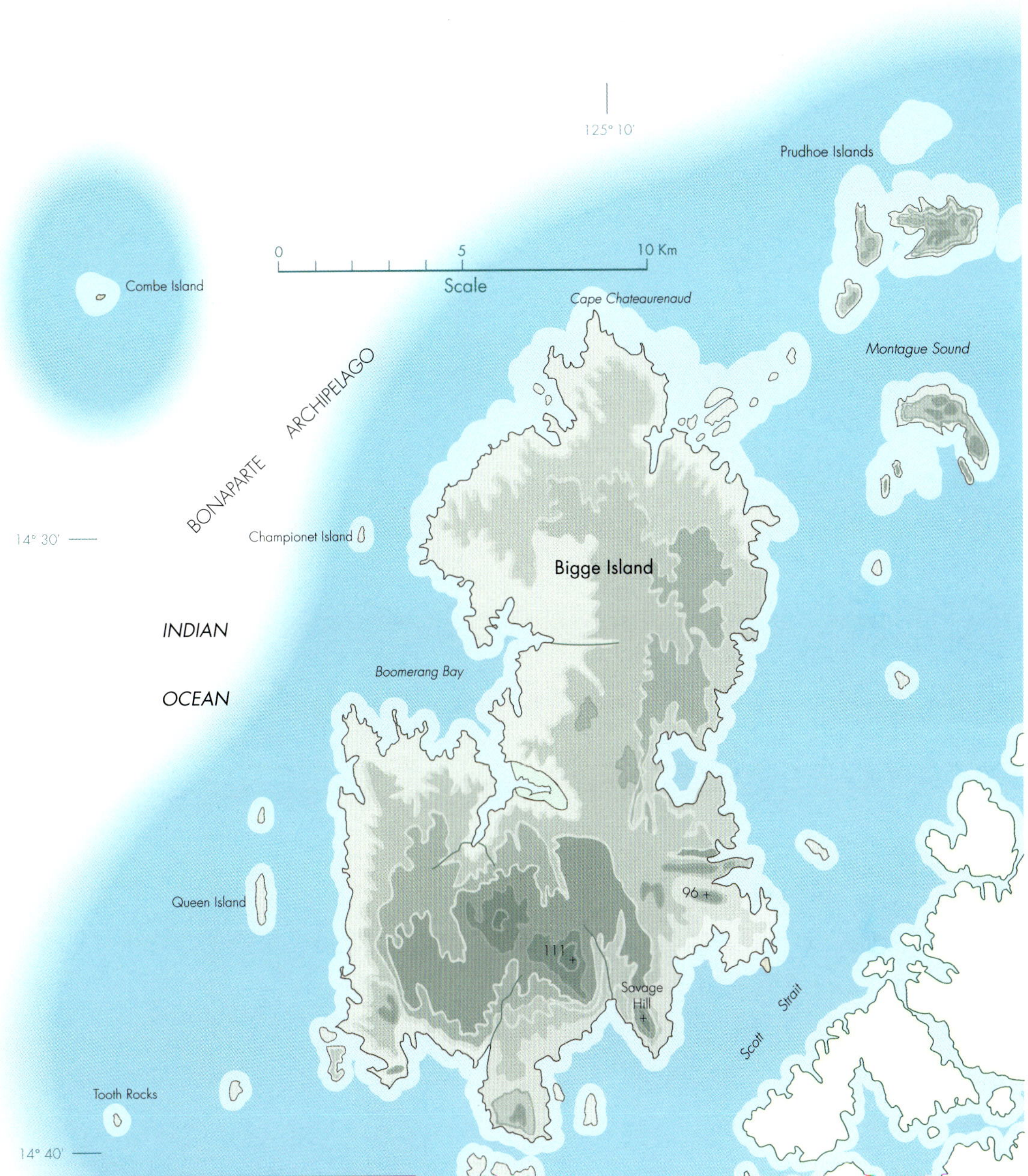

125°10'

Prudhoe Islands

0   5   10 Km
Scale

Cape Chateaurenaud

Montague Sound

BONAPARTE ARCHIPELAGO

14° 30'

Championet Island

Bigge Island

INDIAN

Boomerang Bay

OCEAN

Queen Island

96 +

111 +

Savage Hill

Scott Strait

Tooth Rocks

14° 40'

Combe Island

Admiralty Gulf
Aboriginal Reserve

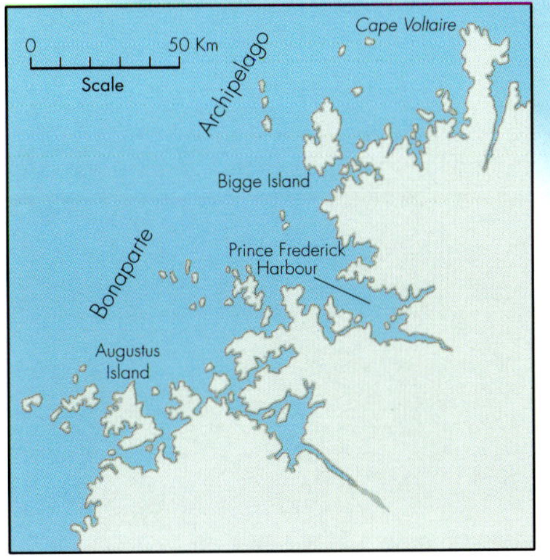

0   50 Km
Scale

Cape Voltaire

Archipelago

Bigge Island

Bonaparte

Prince Frederick
Harbour

Augustus
Island

WESSEL ISLANDS

The End of the Line

A south-easterly blustered in from the Gulf of Carpentaria across the first scrap of land for more than 600 kilometres. Striking an abrupt sea-cliff, the updraft sent several White-breasted Sea-eagles (*Haliaeetus leucogaster*) soaring high. They passed in silent, easy spirals over the windward shore of Marchinbar Island, largest and longest in the Wessel chain. Each bird patrolled its own portion of coast, eagle eye on the upwelling beyond a broken margin of surf, ready to swoop on a Coral Trout (*Plectroproma* spp.) or the silver flash of a Trevally (*Caranx* spp.).

Like a disjointed finger, 60 kilometre-long Marchinbar points north-east from Arnhem Land, far into the Arafura Sea. Immediately to its north, Cape Wessel on the barely detached Rambija Island marks the very end of a chain of more than 20 islands. At 11° South, it lies close to the latitude of Cape York on the opposite side of the gulf.

The three main islands in the Wessel chain, Marchinbar, Guluwuru and Raragala, run for 105 kilometres in a long, shallow arc, yet each is generally no more than three or four kilometres wide, and in places considerably narrower. They are separated from each other by the narrowest of channels. Gugari Rip or The Hole in the Wall is an appropriately named 50 metre-wide slot between Guluwuru and Raragala Islands, navigable only by small vessels at slack tide. Apart from being on the migration route for some bird species, they are not stepping stones to anywhere. In a real and positive sense these islands are at the "end of the line", cut off in time and place from a changing and ever degrading world.

Inclined Plains

The Wessels, and parallel chains to the south-east (the Bromby and English Companys Islands) were separated from continental Arnhem Land in very recent geological time. Their current insularity dates from the end of the Pleistocene Period, when sea levels began to rise after the last protracted episode of world-wide glaciation. Surrounding waters are shallow, so the islands may have been parted from Arnhem Land for only 6–10,000 years. By contrast, their constituent rocks are enormously old — of Precambrian antiquity.

At least 500 million years ago, the Marchinbar Formation was laid down on a gradually subsiding sea floor in deposits of sandstone up to 250 metres thick. More resistant than underlying rocks, these inclined beds now outcrop as the Wessel Islands. The equivalent mainland terrain occurs in the dissected hilly country of northern Arnhem Land draining to the Arafura Sea. Laid back as inclined strata known as cuestas, the islands dip gently, from the abrupt cliffs fronting the gulf in the east, to beaches and low rocky shores in the west.[1]

It was late August and Arnhem Land was firmly in the grasp of the dry season. The near constant, humid trade winds assaulting Marchinbar's eastern coast were likely to deliver only an occasional teasing shower over the next few months. Come the summer monsoon, the winds back to the north-west sweeping lightning-laced sheets of rain over the islands, and out across the gulf.

Coasts of Hope

To the trade winds the Wessel Islands present an easily assailed hurdle. But to the first Europeans this coast was a near unbroken lee shore, a most unwelcoming sight. In April 1623, two Dutch ships set sail from the Moluccan island of Amboina in search of a passage around the southern coast of New Guinea. But instead, the *Pera* and the *Arnhem* unwittingly probed deep into the Gulf of Carpentaria. The two were separated near the south-east corner of the gulf. The *Pera* retraced its outward

journey, while the *Arnhem*, commanded by Willem Joosten van Colster sailed north-west, falling in with the mainland at Cape Grey. Van Colster then headed north and reached "De Caep quade hoop" (Cape of Bad Hope), the weather-side of Marchinbar Island — or as he called it, Het Ejlandt Speult.[2] Unable to continue north along Marchinbar's lee shore, the *Arnhem* turned to the south-west and eventually found a passage through "Het gat de goede Hoop" (Opening of Good Hope) — the Cumberland Strait. From there van Colster sailed for the Dutch outpost on the island of Banda.

A further exploratory fleet sailed from Banda in April 1636, but the voyage of the *Cleen Amsterdam* and *Wesel* was largely unsuccessful. The expedition's commander, Gerrit Thomasz Pool, was killed by Papuans. Then after sailing south and sighting Melville Island and the Coburg Peninsula, the ships retreated, thwarted by "very strong easterly gales".[3]

In 1644, on his second great voyage of exploration, Abel Janszoon Tasman skirted the entire coast of the Gulf of Carpentaria, passed under the lee shore of the Wessel Islands and made the first rounding of Cape Wessel. After becoming embayed near Elcho Island, Tasman's fleet of three ships, the *Limmen*, *Zeemeuw* and *Bracq*, proceeded west, eventually sailing as far as North West Cape. Tasman managed to chart almost a third of the Australian coast but unfortunately any detailed impressions have been lost along with his journals.

The earliest surviving account of the Arnhem Land coast is that of Matthew Flinders in HMS *Investigator* between February and March 1803. After thoroughly charting much of the gulf under extremely trying conditions and in an unseaworthy ship, Flinders moved on through parallel island chains, naming in turn the Bromby and English Companys.

In early March, he reached the Cunningham Islands at the southern end of a further group, noting "A third chain of islands commences here ... [and] is doubtless what is marked in the Dutch chart as one long island and ... called Wessel's Eylandt; which I rename with a slight modification, calling them Wessel's Islands ..."[4] At that point, he "judged it imprudent to

continue the investigation longer".[5] He and his men were sick with scurvy and the *Investigator*, unsound when it left England in 1801, was by then quite rotten. Flinders headed for Koepang in Timor seeking urgent repairs. The survey was not resumed. Instead, the patched-up *Investigator* was sailed in haste back to Sydney via the west coast of Australia.

Frustration, misadventure and finally 10 years in a French prison on Mauritius followed as Flinders attempted to reach England, and there replace the *Investigator*. After first surviving shipwreck in the Coral Sea, he again encountered the Wessel Islands during a second attempt to sail home. The stay was brief. Finding a passage between Marchinbar and Guluwuru Islands (van Colster's Het gat de goede Hoop), Flinders headed for Timor,

naming the narrow strait full of "ripplings and whirlpools" after the schooner *Cumberland* in which he now sailed.[6]

Phillip Parker King followed Flinders in 1819, but made only cursory observations of the Wessel Islands, being constantly troubled by strong and unpredictable tidal races. In the 178 years since, these slivers of Arnhem Land have remained very much as they were in the era of European exploration.

Island Life

Our flight by light aircraft from Darwin to Marchinbar Island was smooth but unfortunately much of the view en route was obscured by cloud. Descending in the late

Above: The lagoon behind Jensen Bay is the only permanent fresh water on the Wessel Islands. About 50 hectares in extent, its margins support typical aquatic plants such as Water Lilies (Nymphaea violacea) and White Snowflake Lilies (Nymphoides indica).

Opposite: The Climbing Fern (Stenochlaena palustris) growing in rainforest behind Jensen Bay on Marchinbar Island. Wet rainforest is limited in extent, mostly occurring in scattered patches on Marchinbar and on Raragala Island. The lagoon behind Jensen Bay is fringed by a rainforest patch of about 30 hectares.

James Barripang's son Richard meets success fishing from the shore at Jensen Bay. Marine resources are of enormous significance to the Aboriginal people of the Wessel Islands. For the 10 or so groups with ownership aspirations over parts of the island chain, marine characters and totems are perhaps as important as the features of the islands themselves in the determination of clan estates.
Rob Jung

afternoon to Jensen Bay at the northern end of the island, the sudden glimpse of such a narrow slice of land in the open ocean was both exhilarating and alarming. It was soon apparent that Marchinbar's only airstrip was a faint brown smudge running half way across its pinched flanks. At last safely landed but still composing ourselves after 400 metres of rough and stony uphill airstrip, we clambered out to meet a small group of Aboriginal people gathered for our arrival.

The Wessel Islands are Aboriginal land. We were privileged to visit Marchinbar with the kind permission of James Barripang, head of a small extended family group belonging to the Guulpa clan of North East Arnhem Land's Yolngu people. Maintaining their traditional way of life from an outstation camp at Jensen Bay, Barripang, his wife, children, grandchildren and several old women, were camped for the dry season on the sand close to the airstrip. They were also 100 metres or so from a magnificent rainforest-fringed lagoon, secreted away behind the foredunes. While in residence on Marchinbar, the men hunt and fish from the shore or from Barripang's runabout. Each day the women collect bush tucker and shellfish. Both men and women collect turtle eggs at nesting time. During the summer monsoon, Barripang and his family often retire with other outstation groups to the 1,200-strong Galiwinku Community on Elcho

Island.

We established our own camp beneath a grove of Coastal She-oaks (*Casuarina equisetifolia*), a few minutes' walk along the beach to the south of Barripang's. The path to our tents struck up through foredunes clad in Beach Spinifex (*Spinifex longifolius*) and dense thickets of shiny apple-green leaved *Scaevola sericea*, both ubiquitous strand species across tropical Australia. Patches of magenta Mulla Mullas (*Gomphrena canescens*), papery heads waving in the breeze, led onto a carpet of casuarina needles embroidered with runners of pink, pea-flowered *Canavalia maritima*. To the west, the day's remnants were swallowed by a placid Arafura Sea. In the soft tropical darkness a small flame flickered among the trees further along the bay. Scented smoke wafted past, entwined with the murmur of women's voices — half song, half talk. Total calm.

Bowerbirds and Bandicoots

The dunes behind our camp backed onto straggly swatches of vine thicket that had stabilised broad expanses of sand. Common tropical trees such as the semi-deciduous Celtis (*Celtis philippinensis*) were interwoven with tangles of creepers and thorny vines. These included Supple Jack (*Flagellaria indica*), climbing on distinctive tendril-tipped leaves and Crabs Eye Vine (*Abrus precatorius*) with its wizened pods of brilliant red but highly toxic seeds. From somewhere among this dense confusion, the Peaceful Dove's (*Geopelia placida*) cooing and the bizarre hissing voice of that consummate avian architect, the Great Bowerbird (*Chlamydera nuchalis*) could be heard throughout the day. Their repetitive calls became almost too familiar as they combined in melody above the sighing continuo of the casuarinas.

One hundred and thirty-nine bird species have so far been recorded in the Wessels, reflecting a mix of sea, shore and bush habitats. But the chain is not a major seabird breeding ground nor is it a foremost stopover on migratory itineraries. Also, as is common with island chains, the sample of bird species and other vertebrates is impoverished compared to the adjacent mainland. As a corollary, though, individual populations may occupy a wider range of habitats and are often more abundant than could otherwise be expected.

On Marchinbar, Raragala and Guluwuru Islands this abundance is exemplified by the Short-eared Rock-wallaby (*Petrogale brachyotis*). Ambling over dissected pavements and rocky broken headlands, we often disturbed unsuspecting wallabies. In an instant they could scamper off down tortuous alleyways. Occasionally stopping under the shade of a rocky ledge, they would nervously watch the intruder into their world before suddenly bounding off, dissolving in the landscape for good. The Short-eared Rock-wallaby exists as several racial variants across monsoonal Australia from the Kimberley to the Gulf of Carpentaria. The Wessel wallabies are appreciably smaller than their continental cousins and may yet be confirmed as a new species.[7]

Seventeen mammals have been noted on the Wessel Islands. With the exception of dogs (part-mongrel, part-dingo) there are no feral species, affording small marsupials such as the carnivorous Sandstone Antechinus (*Pseudantechinus bilarni*) a reasonably untroubled existence.[8]

The importance of these islands as a place of refuge was made clear by a 1993 biological survey carried out by the Parks and Wildlife Commission. In the course of field work, dog droppings were collected on Marchinbar Island and later analysed for dietary clues. Unexpectedly, they were found to contain hair from the Golden Bandicoot (*Isoodon auratus*), a once common mammal that lingered across much of the Northern Territory and extensive parts of the arid and north-western Australia until the 1930s.[9] The Golden Bandicoot's only mainland strongholds are now the most rugged and remote parts of the northern Kimberley. Its island stocks are more secure if somewhat limited, being the most common mammal on Barrow Island off the Pilbara coast. The discovery of a seemingly numerous population on Marchinbar Island extends both its range and continued viability as a species.

Coins in the Sand

Barripang sat cross-legged on the sand. As we talked he kept half an eye on his son Richard, standing at the water's edge. Richard suddenly stooped, every muscle taut. He took two or three slow and high-stepping strides, then hurled his spear about 10 metres out into the shallows. It quivered as it hit its mark. Richard lunged instinctively, snatching the spear with fish attached. In an instant he had buried the fish in the sand for later retrieval, and strode off along the beach, spear ready at hand.

*Golden Bandicoot, Marchinbar Island. The population of Golden Bandicoots has been estimated at around 1,400 individuals on Marchinbar, its only surviving locale in the Northern Territory. A second mammal, the Golden-backed Tree-rat (*Mesembriomys macrurus*), is now extremely rare or possibly extinct in the Territory. It is thought to still survive on Marchinbar Island. Kym Brennan, NT Parks and Wildlife Commission*

Above: Japanese lugger crewman, Marchinbar Island. In 1907, Indonesians were banned from trepang gathering in northern waters but Japanese luggers continued to work the Arnhem Land coast until the late 1930s. In rock shelters, Macassan prahus and Japanese pearling luggers sit side-by-side on ceilings with images of Dingoes, crocodiles, stingrays and a host of marine creatures.

Previous pages: Looking along the Gulf of Carpentaria cliff-tops of Marchinbar Island, north-east of Jensen Bay. Large expanses of hummock grassland, dominated by the spinifex *Trioda microstachya, occur on the exposed flanks of Marchinbar and other islands. Rob Jung*

Back with our conversation, Barripang described the importance of the island to his clan. He was welcoming and keen that we should experience something of his island home but also anxious that we keep clear of any burial sites or places of ritual significance when not in his company. To this end we carefully marked our map.

Barripang fussed over his long bamboo pipe. With a stout twig he poked at the cut-down rifle cartridge that served as its bowl. Then after several attempts, he managed to light the contents, his striking curls and grizzled beard soon brushed by wafts of sweet blue smoke. Barripang's pipe was more than a favoured companion. Its distinctive design was a small reminder of the influence of Macassans on the Arnhem Land people.

For centuries, Macassans had travelled each year to northern Australia from Sulawesi in the Indonesian archipelago. They sailed ahead of the monsoon to numerous sites, from the

Lacepede Islands off the Kimberley coast, east to the bottom of the Gulf of Carpentaria. Setting up permanent camps on islands and mainland beaches, they gathered and cured trepang (sea cucumbers or bêche-de-mer) before returning fully laden on the south-east trade winds to their home port of Macassar. From Macassar these marine delicacies were shipped to eager Chinese markets.

As far as is known, the Macassans did not operate from the Wessels, although the island chain lay across the route of their annual migration. Flinders, having already encountered their shore stations in the Sir Edward Pellew Group and in Caledon Bay on the Arnhem Land coast, ran headlong into the south-bound fleet among the English Companys Islands in February, 1803: "The chief of the six prows was a short elderly man named *Pobassoo*; he said there were upon the coast, in different division, sixty prows ... belonging to the Rajah of Boni, and carrying one thousand men, [that] had left Macassar with the north-west monsoon, two months before ..."[10]

On his return the following October in the *Cumberland*, Flinders landed on Marchinbar Island and found evidence of Macassan visits, intended or otherwise. He also encountered the island's inhabitants: "Four or five Indians made their appearance, but as we advanced they retired; and I therefore left them to themselves ... A Malay Prow had been thrown on the beach, and whilst the boat's crew were busied in cutting up the wreck for fuel, the Indians approached gradually, and a friendly intercourse took place ..."[11]

The relationship between the Macassans and the Arnhem Land people was close, if not always harmonious. As well as trade in artifacts, sexual favours were exchanged. Aborigines occasionally returned to Macassar, and sometimes married there. Macassan words and stories found their way into Aboriginal language, art and belief. The Macassans would undoubtedly have known of the freshwater lagoon behind Jensen Bay. When a radar station operated on Marchinbar Island during World War II, antique Dutch and African coins were found close by, pointing to Macassan visits at some time in the distant past.

Clan Estates

While the Macassans have come and gone over the last three or more centuries, the Aboriginal association with Arnhem Land is inordinately older, dating back perhaps 50,000 years.[12] Over long periods of that residency, the islands would have been low hills and ridges on the wide coastal plain linking northern Australia with New Guinea. At these times the people were probably lowland and coastal dwellers, retiring to higher ground only as sea levels rose.

From the time of their insularity, it is uncertain how much use was made of the Wessel Islands by Barripang's forebears and other groups. Because of its reliable freshwater, Marchinbar was probably the only island to be continuously occupied. The others would have been visited by small groups as seasons and needs dictated. In this century, despite disruption from Japanese pearl and trepang poachers, Christian missions and World War II, Marchinbar has supported a small but more or less permanent population.

The eldest son of Harry Djinjgulul (Tchinggalu), Barripang was born on Marchinbar and inherited the custodianship of the Guulpa estates on the islands and surrounding seas from his father.[13] Fundamentally, his people are of the sea: they call themselves "salt water people" and are skilled hunters of Dugong, turtles and fish. In traditional times they travelled long distances in bark canoes, later replaced by dugout craft obtained from the Macassans. As recently as Barripang's youth, trees were still cut and fashioned for canoes by the Wessel Islanders.

In matters spiritual, Yolngu dreamings are of the sea and often concern creatures such as the shark, sting-ray and crocodile. They unite all clans and extend the dreaming trails of inland people far out to sea. Their totems and ancestral beings are intertwined with the landscape history and natural processes of the coast, islands and the sea. The turtle, for instance, swims the length of the Wessels each year. As she surfaces and exhales, her breath forms clouds that bring calm and safe fishing to the island chain.[14] Barripang's own totem, the Trevally, is, like the man, part of the sea — a spectacular fish often seen flashing through the shallows in mercurial schools.

Just as the sea provides both physical and spiritual sustenance to the Guulpa and other Yolngu peoples, so the islands and coastal waters are vital living and breeding places for many of the animals from their pantheon. Four species of turtle nest on a scattering of sandy Wessel beaches. Among them is the Olive Ridley (*Lepidochelys olivacea*), the smallest, and in Australian waters perhaps the rarest of sea turtles. Its breeding stronghold is North West Arnhem Land.

Other marine creatures enjoy both salt and freshwater habitats. Strolling by Jensen Bay's gentle curve, we often sighted fresh tracks running from the shore towards the dunes and ultimately the lagoon. They were unmistakably those of large Estuarine Crocodiles (*Crocodylus porosus*).

Like the marine resources of all Australian

waters, those of the Arafura Sea (Manbuyngaga ga Rulyapa to the Yolngu) are under pressure. Increased commercial fishing, oil exploration and shipping movements, may place these northern waters at risk. The delineation of Australia's Exclusive Economic Zone, by fixing the northern boundary with Indonesia, has created problems for the Yolngu, effectively severing areas of their dreaming paths and sacred sites in the sea. The Yolngu, together with the Northern Land Council, have recently formulated an Indigenous Marine Protection Strategy for Manbuyngaga ga Rulyapa that seeks to reassert their rightful position in the management of the region's marine resources and protect their traditional associations with the sea.[14]

Chain of Refuge

We sought midday shade under a jutting sandstone overhang, blessed with a panoramic view to the north. Past fractured slabs imprinted with ripples from a primordial Arafura shore, a saddle sloped gently to the sea-cliff then down to the lagoon. On its northern side, rough-hewn sheets of stone and hummocks of spinifex gave way to bare patches of rust-red laterite. This earthy skin signalled one of Marchinbar's bauxite deposits. Some 10 million tonnes have been mapped, spread between seven deposits. The Marchinbar reserves are considered economic but are of low yield and have not yet been exploited.

Out on the gulf, a ship steamed north for Cape Wessel, probably with a cargo of Groote Eylandt manganese secreted within its holds. Unlike its Dutch predecessors, it appeared totally untroubled by Marchinbar's lee shore. My thoughts then turned to the idea of a similar vessel engorging bauxite at some yet to be developed "Port Marchinbar". If that prospect ever arose, should it too be considered an "opening of good hope"? Ultimately, that must be decided by those like Barripang with an intimate connection to Manbuyngaga ga Rulyapa. But for such islands barely altered since the Dutch sailed past, let alone the previous 6,000 years, may they always be valued and protected as a link in Australia's greater island chain. A chain of precious refuge in a shifting sea.

1   A cuesta is defined as a long low ridge with a steep scarp slope and gentle back slope, formed by the differential erosion of strata of varying hardness.

2   Named for the Governor of Amboina, Herman van Speult, sponsor of the expedition. G. Schilder, *Australia Unveiled*, Theatrum Orbis Terrarum, Amsterdam, 1976, pp. 94–5.

3   *Ibid.*, p. 136.

4   Matthew Flinders, *A Voyage to Terra Australis*, G. and W. Nicol, London, 1814 (facsimile edn, Libraries Bd of SA, 1966), vol. 2, p. 246.

       Flinders had previously noted: "The Dutch chart contains an island of great extent, lying off this part of the North Coast ... in some authors it bears [the name of] Wessel's or Wezel's Eylandt, probably from the vessel which discovered Arnhem's Land in 1636." *Ibid.*, p. 234.

5   *Ibid.*, p. 247.

6   *Ibid.*, p. 346.

7   Unpublished information supplied by the Conservation Commission of the Northern Territory.

8   The Sandstone Antechinus is restricted almost totally to Western Arnhem Land.

9   Unfortunately a member of a group of native mammals within a critical weight range, the Golden Bandicoot failed to survive the advent of competition from giant herbivores, namely beef cattle. Much of its decline can also be attributed to predation by feral cats. See A. A. Burbidge and N. L. McKenzie, 'Patterns in the modern decline of Western Australia's vertebrate fauna: causes and conservation implications', *Biological Conservation*, vol. 50, 1989.

10   Flinders, *op.cit.*, pp. 229–30.

11   *Ibid.*, p. 347.

12   Dates of 45,000 +/- 9,000 years obtained in Arnhem Land are still the subject of debate among archaeologists. See R. G. Jones, R. Jones and M. A. Smith, 'Thermoluminescence dating of a 50,000 year-old human occupation site in Northern Australia', *Nature*, vol. 345, 1990.

13   The Guulpa are rival claimants over the Wessel Islands with the Warramiri clan and others. Their claims are based on continuous presence, sacred ties and custodianship. The discovery of bauxite in 1946 eventually led to acrimony between the clans in the 1970s when prospecting authorities were applied for. I. McIntosh, *The Whale and the Cross, Conversations with David Burramurra MBE*, Historical Society of the Northern Territory, Darwin, 1994.

14   P. Sutton (ed.), *Dreamings, The Art of Aboriginal Australia*, Viking, Melbourne, 1988, p. 55.

15   *The Maritime Legislation Amendment Act 1994* sets the border between the two countries mid-way between the Australian continent and the Indonesian Archipelago. See The Ginytjirrang Mala, *An Indigenous Marine Protection Strategy for Manbuynga ga Rulyapa or the Arafura Sea*, Northern Land Council and Ocean Rescue 2000, 1994.

WESSEL ISLANDS

LOCATION

Alger Island: 11° 53' S, 135° 57' E.

Cape Wessel: 11° 00' S, 136° 45' E.

A chain of more than 20 islands extending north-east from the Napier Peninsula, North East Arnhem Land. The closest town is Nhulunbuy (Gove).

AREA

Marchinbar Island: 21,190 ha
Raragala Island: 9,232 ha

CLIMATE

Tropical, monsoonal

STATUS

Aboriginal land. Managed by the Northern Land Council.

ACCESS

Strictly no public access without a permit. Refer to the Northern Land Council, Nhulunbuy.

FACILITIES

None

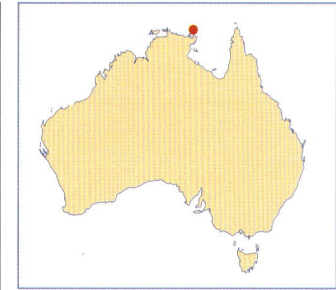

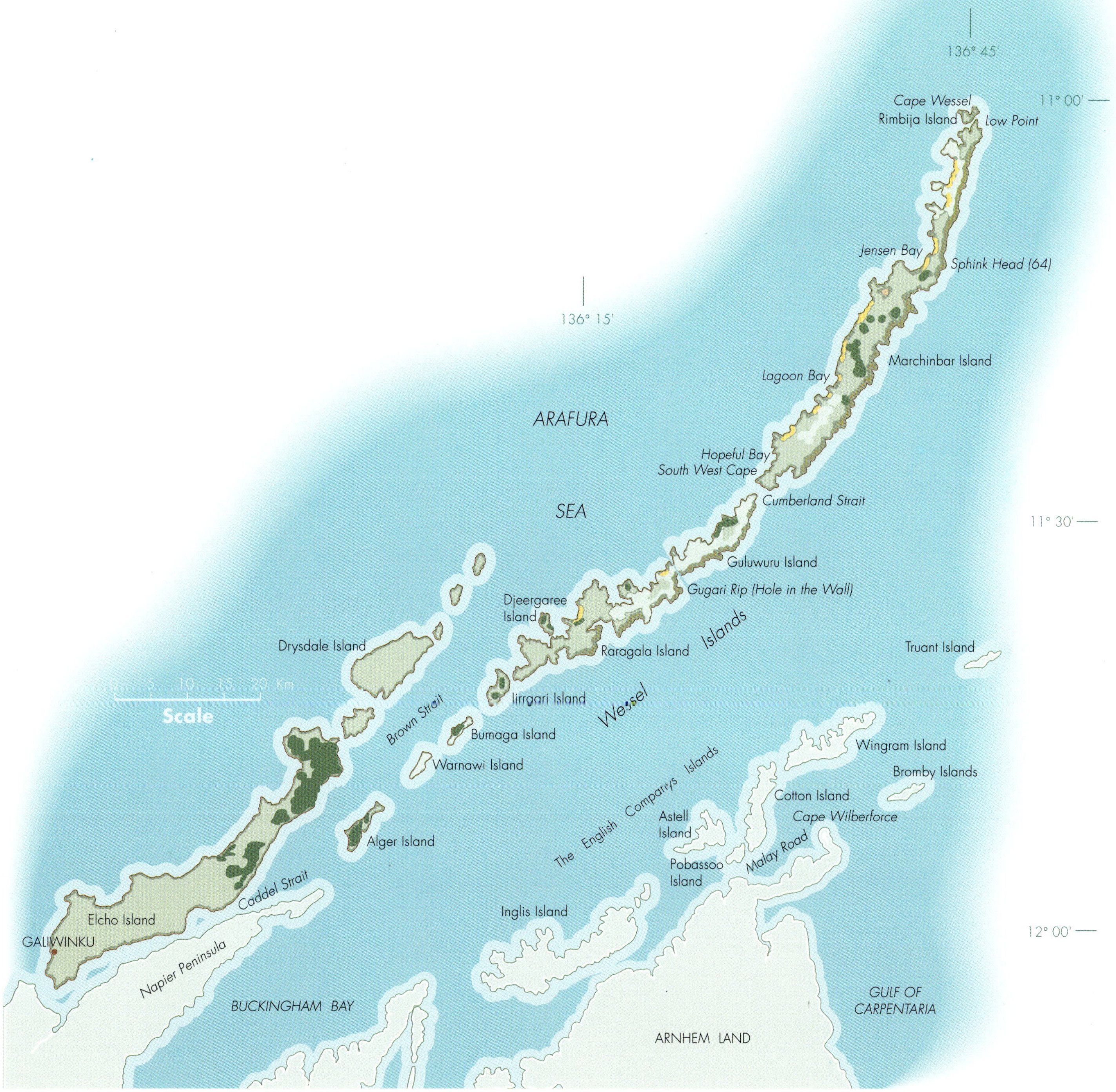

136° 45'

11° 00'

Cape Wessel
Rimbija Island   Low Point

Jensen Bay
Sphink Head (64)

136° 15'

Marchinbar Island

Lagoon Bay

ARAFURA

Hopeful Bay
South West Cape

SEA

Cumberland Strait

11° 30'

Guluwuru Island

Djeergaree Island

Gugari Rip (Hole in the Wall)

Drysdale Island

Raragala Island

Truant Island

Islands

Iirrgari Island

Wessel

Bumaga Island

Warnawi Island

Wingram Island

Scale

0   5   10   15   20 Km

Bromby Islands

Brown Strait

Cotton Island

Cape Wilberforce

The English Company's Islands

Astell Island

Alger Island

Malay Road

Pobassoo Island

Caddel Strait

Elcho Island

Inglis Island

GALIWINKU

Napier Peninsula

BUCKINGHAM BAY

GULF OF CARPENTARIA

ARNHEM LAND

12° 00'

NORTH ISLAND – SIR EDWARD PELLEW GROUP

To the Waters of "Paradice"

Gulf Country. The name conjures up images of wide spaces, flies, heat, dust and cattle — the quintessential Outback. Best seen from the air, flat expanses of woodland merge into the frayed swirls of rivers, creeks and deltas that drain the uplands of the Barkly Tableland. Great tracts of floodplain and tidal reaches, at times up to 40 kilometres wide, stretch from the Roper River in the west, nearly a thousand kilometres to the bottom of the Gulf of Carpentaria at Burketown before turning north for Cape York.

It was early September, the height of the dry season, when we flew into the tiny cattle and fishing town of Borroloola, our staging point for the Sir Edward Pellew Group. Hot desiccating winds from the interior raised flurries of drab-red dust and sent hundreds of Whistling Kites (*Haliastur sphenurus*) soaring on updrafts over the town. It had been blowing hard for days; the wet season and the annual flooding rains were still some months off.

Out on the shallow gulf the breeze was blowing more than 30 knots, raising steep wind-waves and a dangerous sea. The bottom of the gulf is not open to long ocean swells but its unprotected expanse becomes very rough, very quickly.

Thankfully by the next morning the wind had dropped and we were able to leave by swift "shark-cat" for the coast, a journey of 45 kilometres along the McArthur River and the Carrington Channel. Muddy alleys and the occasional Saltwater Crocodile (*Crocodylus porosus*) eventually gave way to the crystalline blue of the gulf.

Our destination was Paradice Bay (Wiyibi) on North Island, the smallest and most distant of the five major islands in the Pellew Group, lying 35 kilometres north-east of the McArthur River mouth. Although the name Paradice commemorates the efforts of an otherwise forgotten naval surveyor, the gentle turquoise waters and sweep of sand backed by groves of Coastal She-oaks (*Casuarina equisetifolia*) justified an altered spelling to describe our idyllic first encounter.[1]

"... They did not discover anything important ..."

The Gulf of Carpentaria coast, enclosing nearly 500,000 square kilometres of sea, was the earliest recorded part of Australia to be explored by Europeans. In 1606, Dutchman Willem Jansz in the *Duyfken*, sailed from the Indonesian archipelago in an exploratory voyage bound for New Guinea. Skirting its southern shores, Jansz turned further south, making landfall on the western coast of the Cape York Peninsula. Several inconclusive voyages followed, before Abel Janszoon Tasman made the first extensive investigations of the gulf in 1644. Also trying to explore the New Guinea coast, Tasman sailed along the western side of Cape York and, like his predecessors, missed the elusive Torres Strait.[2]

Tasman passed the Staaten River, then headed west along unknown shores. Skirting the Wellesley Islands and the Pellew Group, he thought their low, unrelieved profiles to be peninsulas rather than islands. The Dutch were not impressed by the gulf landscape, its inhabitants or commercial potential. Anthony van Diemen, Governor-General of Batavia, assessed Tasman's voyage, stating in correspondence, "[they] ... only found a great spacious gulf or bay ... They did not discover anything important, but only found wretched naked beachcombers without rice or other valuable produce ..."[3]

Further European exploration waited for more than 150 years. Matthew Flinders on HMS *Investigator* surveyed the Gulf of Carpentaria as part of the first circumnavigation of the Australian continent. After negotiating the Torres Strait and following the same course as Tasman, Flinders arrived at a group of islands close to the gulf's south-western corner in

*Above: The sands of Paradice Bay are littered with countless thousand Screw or Turret Shells (*Turritella sp*).*

Opposite: Low tide on the shores of Paradice Bay, North Island. At times in their very recent geological history, the islands of the Pellew Group stood as rocky outliers on a vast sandy plain, several hundred kilometres from the sea. With the last inundation of the Gulf of Carpentaria between 6,500 and 8,000 years ago, these outliers returned to island form.

December 1802. Describing the mainland coast thereabouts as one of "tedious uniformity", his impressions matched those of the Dutch.[4]

While searching for water and wood, he surveyed the islands and their surrounds, honouring a naval colleague when naming the group.[5] He commented in his journal on the elusive local inhabitants but no direct contact was made. Of greater curiosity to Flinders was the puzzling evidence of mysterious visitors:

"Indications of some foreign people having visited this group were almost as numerous and as widely spread as those left by the natives. Besides pieces of earthen jars and trees cut with axes, we found remnants of bamboo lattice work, palm leaves sewed with cotton thread into the form of such hats worn by the Chinese ... but what puzzled me the most was the collection of stones piled together in a line resembling a low wall with short lines running perpendicularly behind into compartments. In each of these were the remnants of a charcoal fire, and all the wood near at hand had been cut down."[6]

Holothuroidea

Flinders had stumbled upon the camps of Macassans, visitors from the island of Sulawesi (Celebes) in the Indonesian archipelago. For the previous hundred years, and possibly many before, up to a thousand men at a time had been making seasonal voyages to northern Australia from their home port of Macassar, a Dutch colonial outpost. The prize of the Macassans, the trepang, for which they braved

dangerous stretches of ocean, consisted of more than 30 edible species of marine invertebrate belonging to the class of Holothuroidea.[7]

Abundant in the shallow waters of South East Asia and northern Australia, trepang was valued as a delicacy by the Chinese. Braised, fried or in soups, as a medicine and even an aphrodisiac, it was shipped from northern Australia to China via Macassar from about 1700 until early this century. It was Australia's first commercial export. The gathering and processing of trepang prior to shipment was an intensive operation. The animals were speared, dredged or collected by divers operating from dug-out canoes. Once ashore, the catch was washed, then boiled in rows of individual cauldrons set between stone grids, as described by Flinders. Next, each animal was gutted and the remaining rubbery muscle boiled again before drying and smoking.

Alfred Searcy, a contemporary observer, encountered "a very charming picture" as he headed for Borroloola aboard the government steamer *Palmerston*: "Four proas were at work, on the shore were the usual smoke-houses, backed up by high hills covered with deep green foliage. Between us and the Malay vessels were a number of the dredging canoes at work, all under sail ..."[8]

The stone lines mentioned by Matthew Flinders can still be found at Webe Point (Wirimbilingaya) on North Island. The site of the old camp is dominated by a grove of Tamarinds (*Tamarindus indica*), a handsome tree introduced to northern Australia by the Macassans. Their camps until recent times were littered with the debris of daily life — broken glass, shards of Chinese porcelain and earthenware pots.

The Place of the Yanyuwa

While the gulf shores prompted contrasting commercial opinions and activities among the Dutch and the Macassans, the Pellew Group has been fundamental to the livelihood of the region's traditional Aboriginal inhabitants, the Yanyuwa, for untold generations. The length of their association with the islands is unknown

Engraving taken from an oil painting by William Westall, artist on the Investigator. *Captioned "View in Sir Edward Pellew's Group, Gulf of Carpentaria" and dated 1812, it depicts Cabbage Tree Cove just south of Red Bluff, so named by Matthew Flinders as "... more abundantly than elsewhere grew a small kind of Cabbage Palm ..." (Livistona inermis). Of particular interest are what appear to be Yanyuwa ceremonial or sacred objects, not present in Westall's site sketch. Flinders' curiosity led him to dismantle sacred stone arrangements on North Island—perhaps the first instance of such desecration in northern Australia. Courtesy of State Library of South Australia, Adelaide.*

but it is certainly a strong one. In their language, the Yanyuwa call themselves Li-anthawirriyarra — a name that speaks of a spiritual genesis associated with the sea.

In the Pellew Group, the Macassans worked on North, Centre and Vanderlin Islands and employed the Yanyuwa in their operations trading canoes, tobacco, and steel axes for turtles and pearl shell. The relationship was long and close. Many Macassan words or their derivatives remain in the Yanyuwa language. Anthropologist Baldwin Spencer described the Yanyuwa people camped near Borroloola in 1901, giving an insight into their trade with the Macassans:

"...a big canoe has come up the river from the coast with about twenty natives in it. It is quite unlike the bark canoes and is simply a great log hollowed out and shaped into a boat. It is quite 30 feet long ... This particular boat was made by the Malays who all come down the coast each year in their Prahus ..."[9]

To the Yanyuwa, the islands were important hunting and food-gathering grounds, and have always been of great spiritual significance. In the rich surrounding seas, the people hunted Dugong, sea turtles and fish. Large shell middens, grindstones, sacred ground ovens (*na-manda*), rock art sites and ritual burials vouch for a traditional life of great continuity. Although no permanent residents remain today on North Island, an Aboriginal outstation is maintained at Webe Point. It is used when ground water supplies permit.

After establishing our own very transient camp in deep shade among drifts of fallen casuarina needles, it was time to explore.

Clumps of Pandanus spiralus *back the sandstone shore pavement near Paradice Bay. To the north, the rocky peninsulas of Cape Pellew and Ross Point with their attendant rock stacks and islets stretch out into the gulf. The next landfall is Groote Eylandt, 130 kilometres away.*

Mindful of the importance of many sites to the Yanyuwa, we kept away from the central part of the island, from Webe Point (Wirimbilingaya) to Red Bluff (Wulibirra). Red Bluff is sacred or secret country (*kurdukurdu*) where the traditional law is still intact. This is the *Yijan* (spirit ancestor's place) of the White-breasted Sea Eagle where the great bird made her nest and performed secret rituals and ceremonies.[10] Red Bluff is accessible only to certain senior men.

Inland and to the south-east from Paradise Bay, the island's highest point, North Hill (Wirdinyjawulaya), rose among weathered sandstone country. We struck a path in its direction, over tottering stone slabs and waving stems of acacia. Up through bristling clumps of grass, the sweet aroma of the spinifex (*Plectracne pungens*) wafted on the breeze.

Despite being only 76 metres high, North Hill commanded good views over the tops of woodland regrowth. The uniform size of the trees on our visit indicated that the island had been hit by a cyclone in the recent past. Distinctive among the stands of young Stringybark (*Eucalyptus tetrodonta*) were scattered Cypress Pine (*Callitris intratropica*) and the showy splotches of orange flowered, Fern-leaved Grevillea (*Grevillea pteridifolia*), a vigorous and common regrowth species in disturbed areas. *Livistona inermis*, a palm of the northern sandstone country, clung to the most

A lone Coral Tree (Erythrina variegata) near Webe Point. An uncommon but highly distinctive small deciduous tree, it is found across parts of tropical Australia and South East Asia.

compromising stony ground, its drooping fan-shaped leaves rattling and gyrating in the breeze.

Rich in plant species, the islands of the Pellew Group show a diversity greater than might be expected for their size. Vanderlin is particularly rich, supporting permanent freshwater wetlands and rainforest patches, as well as woodland, sandplain, dune and mangrove communities.

From North Hill, the seaward extremities of North Island can be seen. The sickled curve of Paradice Bay ends in a tight patch of mangroves backed by salt plains and samphire heath. Powdery dry during our stay, this low-lying expanse would be awash in the wet season, providing a favoured haunt for the Saltwater Crocodile.

We descended directly to the shore, then ambled across ancient multi-hued terraces of sandstone — terracotta, rose and cream — jutting slabs eroded in swirls, incised grooves and delicate honey-combed entablatures. Our path crossed stone of an age difficult to comprehend. As the erosive surface of 100-metre-deep, and 600-million-year-old, Precambrian sediments, these rocks form the primordial skeleton of the Pellew Group. In more recent times, commencing a mere million years past, a shifting coastline moved southward as sea levels rose. The nascent islands became isolated and extensive sediments were deposited in the shallow sea — evident today in the more ephemeral crumbling dune "rock", sands, loam, tidal mud and shell deposits that have accumulated around the group. Dramatic oscillations in the alignment of the coast followed, as successive episodes of world-wide glaciation saw the gentle gradient of the entire gulf drained and flooded periodically.

The Dugong Hunters

A clump of prickly leaved *Pandanus spiralis* clattered its restless answer to the afternoon south-easterly. Our coastal explorations continued. At the southern end of Paradice Bay, on a small beach cordoned by rocks, mounds of sand and eroded tracks above the

tideline, signalled the location of several turtle nests. The Pellew Group is an important habitat for Green (*Chelonia mydas*), Loggerhead (*Caretta caretta*), Hawkesbill (*Eretmochelys imbricata*) and Flatback (*Natator depressus*) turtles. Flatbacks are known to lay their eggs on North and some of the smaller islets throughout the group.

Turtles figure prominently in the stories of the Yanyuwa. The Green Turtle travelled over the northern part of North Island on its dreaming trail, *Malurrba Yijan*, roaming over North Island from Wirimbilingaya (Webe Point) to Langadanga (Cape Pellew).

Just as turtles are important to the Yanyuwa, so the Dugong (*Dugong dugon*) is both game animal, and an important part of the *Yijan*. The dreaming trail of the Dugong Hunters ranges from far in the east to Muluwa (Cape Vanderlin), across the centre of North Island and on to the rest of the Pellew Group, eventually ending at Limmen Bight, in the bottom western corner of the gulf. Before leaving North Island, the Dugong Hunters performed sacred rituals and ceremonies at Winalamba (near Johnson Islet in Watson Inlet). Like Wulibirra, access to Winalamba is restricted by the senior men. Unauthorised entry traditionally resulted in death.

Under the gulf's shallow, light-spangled waters, seagrass flourishes. Herds of Dugong up to 200 strong, browse on these broad carpets as part of a regional population that is among the

Mating Green Turtles. Nesting sites for this species are thought to occur on a number of the Pellew islands. On North Island, Flatback Turtles have been recorded. It is thought that important nesting sites for Flatbacks occur on the smaller sandy islands of the group. Rob Jung

The Fern-leaved Grevillea is common in open woodland of North Island.

Crested Terns. The sandy outliers of the Pellew Group are an important breeding location for Crested and Little Terns. Although widespread, the Little Tern is extremely vulnerable to human disturbance when breeding and is declining in range and number. The isolation of the group provides sanctuary and protection.

five most significant in the world. They are particularly common in the waters adjacent to the mouth of the Wearyan River, south of North Island.

Fruit-doves and Long-haired Rats

Each morning a heavy dew dribbled from the trees, tapping a persistent wakeup call on our tent. Clambering out to a beach cloaked in sea-fog, I was somewhat surprised on one occasion to be greeted by the spectral figure of a lone Black-necked Stork or Jabiru (*Xenorhyncus asiaticus*) as it stepped stiltedly from the mist. A bird I had associated with lagoons and swamps (where it hunts fish, amphibians and crustaceans), this individual seemed somewhat out of place striding an island beach at low tide, so far out to sea. But then the deltas of the gulf coast and even the mudflats and mangrove reaches of some of the islands are probably ideal habitat, and Jabirus have been recorded on South West as well as North Island. Ever wary, it took off as soon as I was spotted, flapping laboriously on striking white and black wings, trailing its long scarlet legs.

The Pellew Group supports a wide array of birds: close to 200 species have been recorded, reflecting a diversity of habitats across the islands. Coastal vine-thickets and mangroves are home to specialised species, including the Emerald Dove (*Chalcophaps indica*), Rose-crowned Fruit-dove (*Ptilinopus regina*) and Torresian Imperial Pigeon (*Ducula spilorrhoa*), Mangrove Golden Whistler (*Pachycephala melanura*), Shining Flycatcher (*Myiagra alecto*) and Mangrove Robin (*Eopsaltria georgiana*), all reliant on their particular restricted habitat for a foothold in the region. Extensive tidal mudflats are the winter resort for large numbers of migratory waders, such as the Greenshank (*Tringa nebularia*) and Ruddy Turnstone (*Arenaria interpres*). Rare and endangered species like the Black-winged Stint (*Himantopus himantopus*) have been sighted among them. The Gulf of Carpentaria mudflats are considered the third most important Australian ground for waders migrating from the northern hemisphere each year. On the

seaward fringe of the group, small rock-stacks and islets are a significant habitat for nesting seabirds. Large congregations of Crested Terns (*Sterna bergii*) are found off North Island and the vulnerable Little Tern (*Sterna albifrons*) is known to breed in the vicinity.

We roamed widely over the rocky peninsula ending in Cape Pellew (Langadanga), an area of special significance to the Yanyuwa. High up, the outlook to the west stretched over a small city of eroded sandstone pinnacles towards Ross Point (Aralwiji) and the open gulf.

Away from the frequent lick of fire, the alleyways and corridors of Cape Pellew's stone citadel are crowded with patches of green. These semi-deciduous vine-thickets make up only about seven per cent of the island's vegetated area but despite their limited extent, they provide vital habitat links for rainforest specialists, such as the Rose-crowned Fruit-dove, with Arnhem Land and Cape York populations.

As each afternoon waned, on the soft sands fronting Cape Pellew, the tracks of the Agile Wallaby (*Macropus agilis*) became noticeable, speaking of much activity during the nights. Often seen grazing in small mobs close to woodland cover at dawn and dusk, they are widespread and common on Vanderlin, North, Centre and South West, as well as the mainland. Agile Wallabies are one of 26 mammal species recorded on the Pellew Group. Four of these have not been recorded on the gulf mainland in recent times. Populations of the Brush-tailed Phascogale (*Phascogale tapoatafa*), Northern Quoll (*Dasyurus hallucatus*), Brush-tailed Rabbit-rat (*Conilurus penicillatus*) and Canefield Rat (*Rattus sordidus*) inhabit the islands, like some bird species, as isolated communities between Arnhem Land and Cape York, or at the geographic limit of their range.

The Pellew Group remains relatively clear of feral animals and introduced weeds. North Island is largely free of both. Cane Toads (*Bufo marinus*) now present on the gulf mainland, have been recorded on South West Island but thankfully appear not to have been successful colonists. Away from feral cats, native rodents such as the Common Rock Rat (*Zyzomys argurus*) and Long-hair Rat (*Rattus villosissimus*) abound on North Island.

A sandstone alleyway on Cape Pellew. Its protected gorges provide refuge for fire-sensitive vine-thickets, which also occur in occasional dense patches on established foredunes. Semi-deciduous vine-thickets are a rare and vulnerable community in the gulf region. With its gleaming creamy limbs, Eucalyptus kombolgiensis *is distinctive and quite common on the rocky sandstone slopes of the Pellew Group. It is otherwise known only from the stone country of Arnhem Land, far to the north-west.*

Exhibiting a richness and diversity now hard-pressed on much of the mainland, the true worth of this coastal cluster of near unspoiled islands must be their persistence as both refuge and link in the changing biogeography of northern Australia.

"A *very important place*"

Our return by road to Darwin was a shock. Days stepping out across the easy terrain of North Island, or that sparkle of a limpid-blue gulf, were quickly beaten into memories by narrow strips of patched tar and the occasional pothole. Leaving Borroloola for Daly Waters, we followed a road for the first 70 kilometres that was anything but an outback track. Wide and straight, with bitumen that would not be out of place on a freeway, it ceased abruptly at the turn-off to the McArthur River project, a huge silver, lead and zinc mine currently being developed by Mount Isa Mines and Japanese investors. From there on, the road reverted to Outback "Territory style" and remained so until we reached the junction with the Stuart Highway at Daly Waters.

The McArthur River Mine is bringing significant change to the region. Port facilities are under construction at Bing Bong on the coast and another is planned for Centre Island as part of an overall strategy for gulf development. But more than the mine and associated developments, the project will bring people in numbers to the gulf. From barges and bulk ore carriers to campers, runabouts and amateur anglers, the solitude of these gentle islands may never be the same again. The Yanyuwa, the Macassans, then the pastoral industry, all have staked their claim or made a mark on the region; only the Dutch remained indifferent. Now mining. How this next wave in the gulf will mix with Dugongs, turtles and migratory waders, only time will reveal.

In 1992 North Island was declared a national park, to be jointly managed by its traditional owners and the Northern Territory Parks and Wildlife Commission. The first reserve in gulf waters, its establishment should be just one among many actions aimed at protecting the cultural and natural worth of what the Yanyuwa rightly know and respect as *walkurra awara nya-mangaji* — a very important place.

---

1   Surgeon-Lieutenant W. E. J. Paradice visited the islands on HMS *Geranium* during a 1923–4 naval survey of the gulf. He made notes on the flora and fauna, collecting fish and reptile specimens.

2   Sailing instructions issued to Tasman in Batavia by Justus Schouten of the Dutch East India Company (VOC) Council, made it "... necessary to investigate whether Nova Guinea is linked to the great Southland or whether it is separated from it by channels between islands that lie between; ..." G. Schilder, *Australia Unveiled*, Theatrum Orbis Terrarum, Amsterdam, 1976, p. 182.

   In 1606 the Spanish navigator Torres, sailing from east to west, successfully negotiated the strait that the Dutch still searched for.

3   Schilder, *op. cit.*, p. 187.

4   M. Flinders, *A Voyage to Terra Australis*, G. and W. Nicol, London, 1814 (facsimile edn Libraries Bd of SA, Adelaide, 1966), vol. 2, p. 159.

5   "I conceive that the great alteration produced in the geography of these parts by our survey, gives authority to apply a name which without prejudice to the original one, should mark the nation by which the survey was made; and in compliment to a distinguished officer of the British navy ... I have called this cluster of islands, Sir Edward Pellew's Group." *Ibid.*, p. 170.

   (Admiral Sir Edward Pellew, 1st Viscount of Exmouth, 1753–1833). Flinders gave names to most of

the islands in the group, including Vanderlin, which he distinguished from the mainland and Cape Vanderlin — "In the old Dutch chart". *Ibid.*, p. 169.

6   *Ibid.*, pp. 172–3.

7   "Trepang" derives from the Malay *teripang* and the alternative, "bêche-de-mer", from the Portuguese *bicho da mar*, meaning "sea worm". At the industry's height more than 300 tonnes of processed trepang were taken back to Macassar each year from northern Australia. This formed a significant part of the quantity of trepang which finally reached China.

8   A. Searcy, *In Australian Tropics*, Kegan Paul, Trench, Trubner & Co., 1907, pp. 120–1. Searcy was Sub-Collector of Customs at Port Darwin in the 1880s. He had many dealings with the Macassan trepangers when the industry in the Northern Territory came under South Australian Government license control. Borroloola became an important riverhead for freight to and from Darwin and the gulf.

9   Baldwin Spencer, Diary 1901–1902, pp. 109–10, as quoted in R. Vanderwal (ed.), *The Aboriginal Photographs of Baldwin Spencer*, Viking O'Neil, 1987, p. 118. Spencer's observations were made at the end of the 1901–2 Gillen and Spencer Expedition from Oodnadatta overland to the Gulf of Carpentaria.

10   *Yijan* can be roughly translated as "dreaming". It can be applied to both the spirit ancestors and the time of creation when these ancestors roamed the land in a multitude of animal, plant and human forms.

NORTH ISLAND – SIR EDWARD PELLEW GROUP

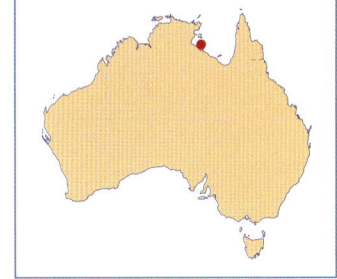

LOCATION

North Island: 15° 35' S, 136° 52' E. One among 7 continental islands and numerous islands and rocks, up to 45 km north of the mouth of the McArthur River, in the Gulf of Carpentaria. The closest town is Borroloola.

AREA

Vanderlin Island: 27,690 ha
Centre Island: 9,222 ha
North Island: 5,421 ha

CLIMATE

Tropical, monsoonal

STATUS

Vanderlin Island and Centre Island are Aboriginal land. **North Island** is the Baranyi National Park, jointly administered by the Parks and Wildlife Commission of the Northern Territory and the Baranyi Aboriginal Corporation through the Local Management Committee (LMC).

ACCESS

Boat access from Borroloola via the McArthur River or the Carrington Channel. Sea conditions can be variable and Saltwater Crocodiles are present. Zoning of North Island intends to restrict levels of access to areas of conservation and cultural significance. Permit required—contact the Parks and Wildlife Commission for details. Public access to North Island only.

FACILITIES

Limited campground and interpretive facilities are being developed on North Island.

ISLANDS OF CAPE YORK AND THE TORRES STRAIT

In the Lee of Bushy Islet

We set sail in darkness from Bushy Islet, confident of catching the tide at the eastern end of Albany Passage near Cape York. It was 3.30 a.m. Dawn was another two hours off and we were still more than 70 kilometres from the cape.

Bushy Islet had proved an uncomfortable but convenient stopover on our passage between Shelburne Bay and the Torres Strait.[1] Tides and headwinds had impeded our progress, dashing any hope of reaching the cape during the previous afternoon. Consequently, this tiny coral island, just seven kilometres from the mainland coast provided welcome but tenuous overnight shelter. As a bonus, our stopover afforded an opportunity to explore a vegetated cay and its attendant reef that differed greatly from any seen further south.

Characterised by colonising mangroves, a leeward sand cay, extensive rubble ridges and sandy reef flat, Bushy Islet is described as a low-wooded island. Approximately 45 of the more than 300 coral islands along the Great Barrier Reef have developed in this way. Formative cays are severe places: nutrients required to spawn plant growth are few, and freshwater catchments small. To sustain mangroves in this setting requires copious rainfall, hence low-wooded islands are only found north of Cairns where annual rains exceeds 1 200 millimetres

Pigeons in the Mangroves

The tide was dead low when we made the short trip by dinghy to the end of Bushy Islet's sweeping tail of sand. It lay at the leeward end of a long, saucer-deep lagoon that in places was completely dry.

The shallows that remained were busy with life. Sloshing towards the mangroves I disturbed a small ray. With the flick of a wing, it shed its cloak of sand and skimmed away for a few metres, settled on the bottom, then quickly

regained its gritty disguise. Also dusted in sand, starfish and bêche-de-mer lay in suspended animation between spreading growths of olive-green soft coral.[2] Like handfuls of multi-fingered rubber gloves, the coral colonies gestured back and forth in their shin-deep pools. Ahead, the wind-ruffled puddles suddenly foamed as two small carpet sharks darted for deeper water.

I soon reached the rubble margins of the drying reef's rampart. Boulder-sized accretions of coral dotted the pavement on the windward south-east side of the island. Cast up by the surf, they formed random lines amid a low wind-shaved hedge of White Mangroves (*Avicennia marina*) and stout-limbed *Aegialitis annulata*. Such uncompromising ground, newly won from the sea, bristled with the *Avicennia*'s oxygen-seeking pencil roots.

In contrast, on the leeward side of the coral rampart, taller trees had gathered in a small mature forest, the looping stilt roots of *Rhizophora stylosa* drilled fast in the sand and mangrove mud. On some islands, *Rhizophora* forms extensive stands, attracting large breeding colonies of Torresian Imperial Pigeons (*Ducula spilorrhoa*) from New Guinea in July and August each year. Noted in 1606 on

Opposite: Forbes Island in Temple Bay near Cape Grenville from the Cape York Peninsula. Forbes is typical of the hundreds of continental or "high" islands within the confines of the Great Barrier Reef. They are merely ancient hilltops and mountaintops that survived rising sea levels 8,500 to 6,500 years ago. Many are surrounded by their own fringing coral reefs.
Ian Brown

Below: A stand of colonising mangroves, Rhizophora stylosa, *on Bushy Islet. At high water the arching prop roots of the mangroves are submerged. As the tide rises small fish are washed over the reef. They graze among the tidal forest, safe from predators. On heavily wooded mangrove islands north of Cairns,* Rhizophora *can form pure stands of trees up to 15 metres high.*

A colony of soft corals in the shallow pools of Bushy Islet's reef flats. Soft corals lack the rigid limestone skeleton of their hard cousins but possess numerous small calcareous spicules within their walls. While not reef-builders, soft corals occupy an important place in the reef's ecology. In some places they can account for about half the reef's living tissue.

Moa (Banks) Island by the Spaniard, Luis Vaez de Torres, the Torresian Imperial Pigeons were possibly the first Australian birds to be sighted by Europeans.

Small Boat Journeys

One hundred and eighty-four years after Torres battled the winds, tides and reefs of the strait that came to carry his name, an open ship's boat with 18 men on board also struggled to find a passage westward. Forcibly cast adrift from HMS *Bounty* by mutineers, William Bligh and his companions had sailed from the Tongan island of Tofua, desperate to reach the safety of Koepang in Timor. After a month on the open sea, they found a passage through the reef and fell in with the Australian coast at Restoration Island near Cape Weymouth on 29 May 1789. Over the following week they headed towards Cape York, catching seabirds and gouging oysters from island shores to supplement their paltry rations. On 2 June, Bligh noted a cape at 11° 18'S and "A long Sandy Isld just above water ... to the Northd & eastward of it ..."[3] Whether this was Bushy Islet is unclear but without the bounty of similar inshore islands, Bligh and his crew would have

certainly perished.

Hard by the top of Australia and far from the madding crowd of Great Barrier Reef tourism, Bushy Islet is a wild and lovely place. Unspectacular and unsung, just one little tile of a great mosaic, it is nonetheless an exquisite example of the diversity of the reef and the islands within its bounds. Turning back to the sand spit and our beached dinghy, the manes of cloud streaming towards the horizon provided a clear reminder that the trade winds were not about to abate. Lines of shell fragments, bleached and broken coral and an assortment of ocean detritus fringed the sandbank like rings of scum on a bath. To the west, the featureless coastal dunes of the peninsula dissolved in the salt-laden afternoon light. The south-easterly was gusting at over 20 knots as we braced ourselves for the soaking ride, back through the swell combing around the reef, to a restless, rolling yacht.

Since leaving Cairns 12 days before, strong south-easterly winds had blown almost continuously, interrupted only by erratic and momentarily violent rain squalls. In the Torres Strait, the trade winds funnelling between northern Australia and the mountains of New Guinea are said to be the strongest and most prolonged in the world. They blow between

Islands of the Endeavour Strait.
Ian Brown

late March and early October, gusting to more than 30 knots on average at least one day each week and consistently range between 15 and 25 knots.[4]

Australia's Singapore

At dawn, the cape was within reach and the protection of the coast-hugging reef began to fade. Wind and swell from the south-east fighting against an outflowing tide made for a confused, sloppy sea. Towards Cape York, the Great Barrier Reef veers away from the coast, tracing the edge of the continental shelf as it heads for Papua. The reef itself peters out in the shallow turbid waters and strong tidal flows of the Torres Strait and is eventually drowned by freshwater flooding from the Fly River.

We headed cautiously for Albany Passage, intent on passing through the gap between the mainland and Albany (Pabaju) Island at slack water. The shortest route to the Torres Strait is also narrow and subject to racing tides. Well-timed though, the yacht slid past two or three shoaling banks and coasted into the passage with the newly turned flow. From open and grassy Fly Point on our port side, barely a few hundred metres across the passage, Albany Island looked impenetrable, its flanks tightly wrapped in wind-tormented vegetation.

It seemed hard to believe that this short waterway was a major port before there was any significant development along the Queensland coast. The gathering and curing of bêche-de-mer for the Chinese market was the first industry in the Torres Strait. From the early 1840s a processing station operated intermittently on Albany Island. The first Governor of Queensland, Sir George Bowen, visited the region in 1862 and proposed the island as a promising site for the "Singapore of Australia", with the potential to grow into a major port-of-call and coaling station for all ships entering or leaving Australia's east coast waters. Before either Townsville or Cairns was founded, the push to the extreme north was on. In part a response to French activity in the region that culminated in the 1853 annexation of New Caledonia, the move also sought to capitalise on the increasing number of ships

plying the Torres Strait between Europe, India and Australia. A settlement developed on either side of the passage. Somerset on the mainland was a bustling colonial outpost for more than a decade.

In the later years of the nineteenth century, the waters around Albany Island were crowded with more than 400 pearling luggers and thousands of divers but the restlessness of the tides flowing through the passage jeopardised its continued life as a port. The Queensland administration, keen to extend its authority further north, quit Somerset in 1877 for Thursday (Wayben) Island, which remains today the commercial and administrative centre of the Torres Strait. About 700 ships now call at Thursday Island each year.

Our course turned north-west into the Adolphus Channel, where we passed Cape York almost without knowing. Continental Australia's unprepossessing northern extremity lay partly hidden behind the more imposing York and Eborac Islands. A broad and seemingly benign stretch of water lay ahead. But in 1891, just six kilometres north-east of Albany Island, shoaling rocks missed by such skilled hydrographers as Phillip Parker King and Owen Stanley claimed the SS *Quetta*, in the region's worst maritime disaster. One hundred and thirty-three lives were loss when the *Quetta* sank in just five minutes.

In today's busy strait, the threat of maritime disaster remains just as real. In 1971 the tanker *Oceanic Grandeur* hit an uncharted rock in the main east–west channel near Thursday Island and spilt over 1,000 tonnes of oil. The already diminished cultured pearl industry was devastated. Just over 200 years before, a small wooden ship had made a more successful passage through these waters.

"… a congeries of islands …"

In 1770, having rounded Cape York in HMS *Endeavour*, James Cook realised that the continent's north-eastern limits had been reached. He duly landed on Possession (Tuined) Island on 22 August and with a simple but sweeping imperial flourish claimed the entire east coast of the continent for King

George III.[5] In one short act the fate of the Torres Strait and Australia was changed for ever.

Cook then negotiated the strait that he named for his little ship and headed hastily for Batavia and urgent repairs. That the *Endeavour* had survived more than 2,000 perilous kilometres of unknown reefs is perhaps the highest testament to the seamanship of Cook, the doyen of navigators. We now had some small inkling as to how Cook must have felt as he finally sailed west.

To the north-west of Endeavour Strait, Cook observed: "... a Congeries of Islands ... which I named Prince of Wales Islands. It is very probable that these Islands extend quite to New-Guinea ... many of them seemed indifferently well cloathed ... and from the smooks we saw, some if not all of them must be inhabited."[6]

The shallow turquoise waters of Endeavour Strait ushered us past Prince of Wales (Muralag) Island to clear sailing and the Gulf of Carpentaria ahead.

Numerous islands and reefs stretch northwards on the boundary between the Arafura and Coral Seas, across the 150 kilometres to the Papuan coast, like sluicegates on the ebb and flow of the tide. The flanks of Muralag, rising to an imposing 246 metres screened a tight cluster of "high" continental landforms that includes Thursday (Wayben) and Horn (Narupay) Islands. Further north a second continental group centres on Moa (Banks) and Badu Islands. Formed of volcanic tuffs and associated granite intrusions more than 300 million years old, they are the very stuff of the mainland, with the same rocks outcropping on the Cape York Peninsula and also in Papua. On the eastern side of the strait, eroded remnants of Pleistocene volcanic activity form the Murray Group (Mer, Waier and Dauar) and Darnley (Erub) and Stephens (Ugar) Islands. And within the central coral maze of patch reefs and channels, sand cays have formed.

The Torres Strait awash is a geographically recent phenomenon. For most of the last 200 million or more years, Australia has been joined to the nascent New Guinea, or the region of its recent bounds, by a broad continental shelf known as the Sahul. Only in the last 8,500–6,500 years, following post-glacial rises in sea level, has the place been transformed from an open forest landscape to a maritime environment of islands and reefs.

Ghosts and Cannibals

Today, the strait supports an indigenous Melanesian population of about 8,000 scattered across 16 islands. A further 15,000 today live on the mainland. It is not known how long the region has been occupied but the people were certainly there when Torres sailed through. The ancestors of today's Islanders came from Papua, probably by single-outrigger canoe like

"An Eye Sketch of Part of New Holland in the Bounty's Launch by Lieut. Wm. Bligh". A remarkably accurate sketch-chart made by William Bligh in 1789 as he sailed in an open boat along the Cape York coast after the seizure of HMS Bounty by mutineers. Bushy Islet and Cairncross Islets to the east, while not plainly shown, are one of a group noted by Bligh with the annotation "Low keys".
Courtesy of National Library of Australia, Canberra

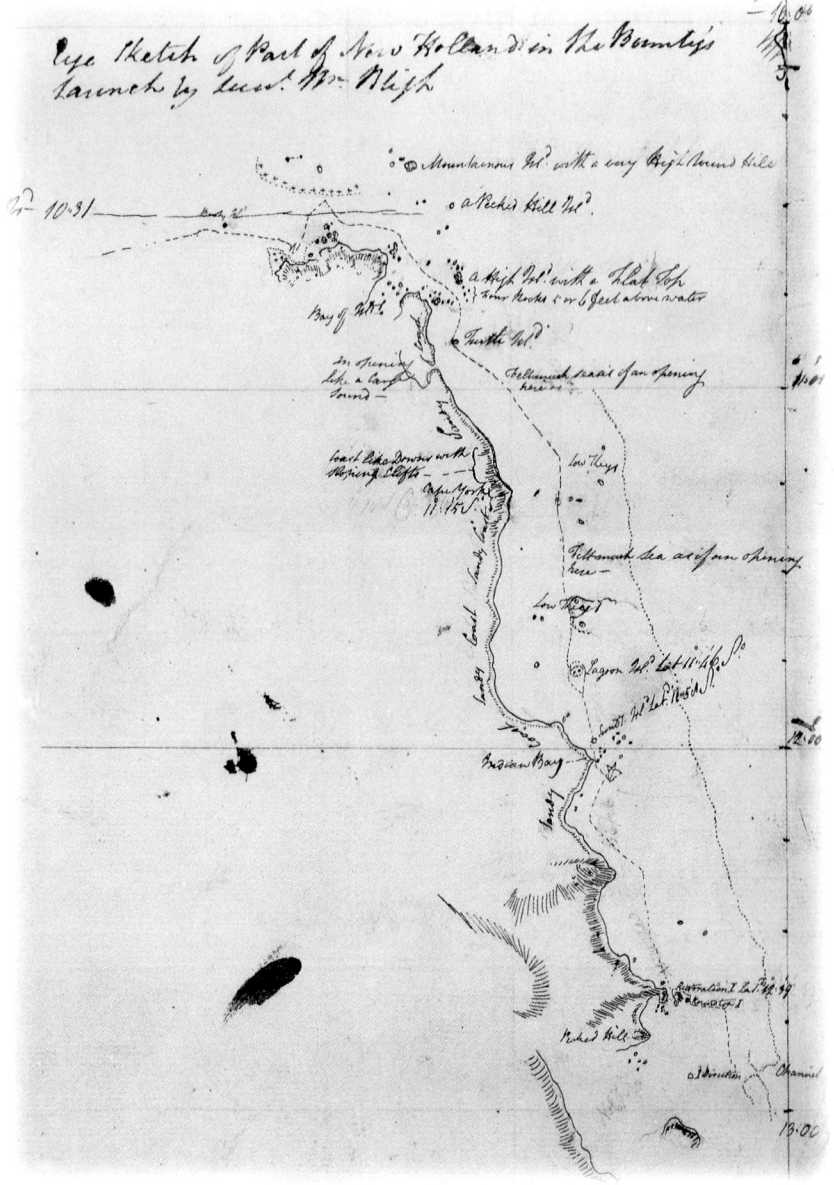

those still used on the Papuan coast today. Successfully colonising the length and breadth of the strait, the people drew both physical and spiritual sustenance from the sea, as they still do. Agriculture, so important to their Papuan cousins, assumed a minor place in the midst of one of the richest marine environments on earth. Simple canoes evolved into massive sail-driven double-outriggers up to six metres long, craft capable of carrying up to 20 warrior crew.

Muralag, among a group of about eight islands close to Cape York, was traditionally occupied by the Kauralgal people. Among the most southerly Melanesian inhabitants of the Torres Strait, the Kauralgal were a semi-nomadic group, perhaps 100-strong.

The hold of the Kauralgal on their territory was probably an uneasy one. Inter-island warfare, headhunting and cannibalism were part of daily life. As a trading link with the rest of the strait and the Aborigines of Cape York with whom they intermarried extensively, the Kauralgal occupancy of the Prince of Wales Group was grudgingly tolerated by their fierce northern neighbours.

With the coming of Europeans, violence did not abate — at least not until missionaries arrived in the 1870s to suppress the cults of old. Early Europeans venturing through the strait were thought to be ghosts, and frequently ambushed and massacred. Luckless castaways often suffered the same fate. Reprisals followed quickly. As more Europeans arrived during the 1860s in search of pearl shell and bêche-de-mer, havoc spread throughout the strait. Within a decade, the Kauralgal threat in the Prince of Wales Group had been "neutralised" and pearling stations established on several islands. The Kauralgal are long gone from Muralag, victims of bloody warfare with Europeans. The

Sunset, looking towards the islands of the Prince of Wales Group from the western side of Cape York.
Ian Brown

pacified survivors were shifted to Hammond (Keriri) Island and then finally to Moa in 1921. A few Kauralgal descendants now also live on Narupay Island.

Muralag is now occupied by Islanders from all parts of the Torres Strait. It has seen cattle grazed on its majestic slopes while deer and horses run wild. With the only reliable water supply in the Torres Strait, a dam, town and tourist resort have all been proposed in the recent past. Within the possibility of such change, elders and leaders among the remaining Kauralgal have sought to identify and protect their sacred sites on Muralag and surrounding islands.

Unlike many mainland Aborigines and the Kauralgal, most Islanders have not been dispossessed of their estates. From the distant Murray Islands, Eddie Koiki Mabo and four other Meriam people waged a 10-year battle for recognition of their inalienable tenure over traditional lands. The High Court "Mabo Decision" of 1992 upheld these rights and created the precedent for the recognition of native title throughout Australia, but it did not include customary marine tenure. As at June 1996, some 70 applications covering both land and sea rights had been filed under the Native Title Act in the Torres Strait. Determinations are yet to be made.

The 1978 Torres Strait Treaty between Australia and Papua New Guinea sought to safeguard Islanders' livelihoods won from the sea and provide freedom of movement between both sides of what is in reality a border between nations.[7] The treaty has not been the success that Islanders craved. With two national, as well as state and local governments involved, Islanders often feel sidelined by bureaucracy. An inevitable push for greater control and eventual autonomy has begun. But many of the problems facing the strait do not fit within lines on a map. Pollution from mining discharges in Papua New Guinea, shipping accidents, resource depletion and the threat of sea level change due to global warming, all loom large on the Torres Strait's horizon.

Muralag Island slid astern, its graceful silhouette fading in the humid afternoon haze. The view would not have been different from that seen by Cook, gazing back in satisfaction after escaping the clutches of the Great Barrier Reef and the Torres Strait.

We began to feel a new motion in the seas sliding past our keel. Wind-waves borne on the south-east trades pushing across the open gulf were now merging with the currents that poured back and forth through the strait. Three days of rocking and rolling lay ahead to the north-east coast of Arnhem Land. The islands drifted peacefully behind. Along with their turbulent past and uncertain future, they were soon gone from view. For the time being we slipped silently into a world without end, a world without land.

1   Not to be confused with Bushy (Redbill) Island near Mackay.

2.   Bêche-de-mer is also known as "trepang". Early accounts of the Queensland industry refer to "bêche-de-mer".See chapters 6 and 7.

3   J. Bach (ed.), *The Bligh Notebook 'Rough account — Lieutenant Wm Bligh's voyage in the Bounty's Launch from the ship to Tofua & from thence to Timor'*, Transcription and facsimile, Allen and Unwin, Sydney, 1987, p. 137.

4   Heating of the tropical air mass in the Northern Hemisphere causes massive convection currents that draw in air across the Equator from the south. The Earth's rotational spin does the rest, skewing the direction of the air flow's origin to the south-east. When the meteorological Equator moves south in the southern summer, the process is reversed.

5   Cook proclaimed on Possession Island:
    "...the Eastern coast from the Latitude of 38° South down to this place I am confident was never seen or viseted by any European before us ... I now once more hoisted English Coulers and in the name of his Majesty King George the Third took possession of the whole Eastern Coast from the above Latitude down to this place by the name of *New South Wales*, together with all the Bays, Harbours Rivers and Islands ... after which we fired three volleys of small Arms..."
    J. C. Beaglehole (ed.) *The Journals of Captain James Cook on His Voyages of Discovery*, vol. 1, Hakluyt Society, Cambridge, 1955, pp. 387–8.

6   *Ibid.*, pp. 390–1.

7   The border between the two countries is drawn close to the Papuan coast. It was fixed by the Queensland Government in the 1870s without regard to international convention. Queensland sought to assert control over shipping movements, fishing grounds and the perceived German colonial threat in the region at the time. See J. Singe, *The Torres Strait, People and History*, University of Queensland Press, St Lucia, 1989.

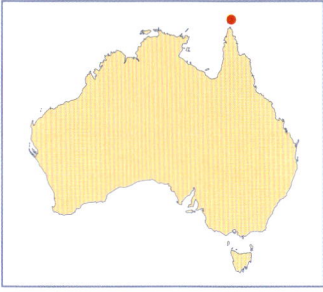

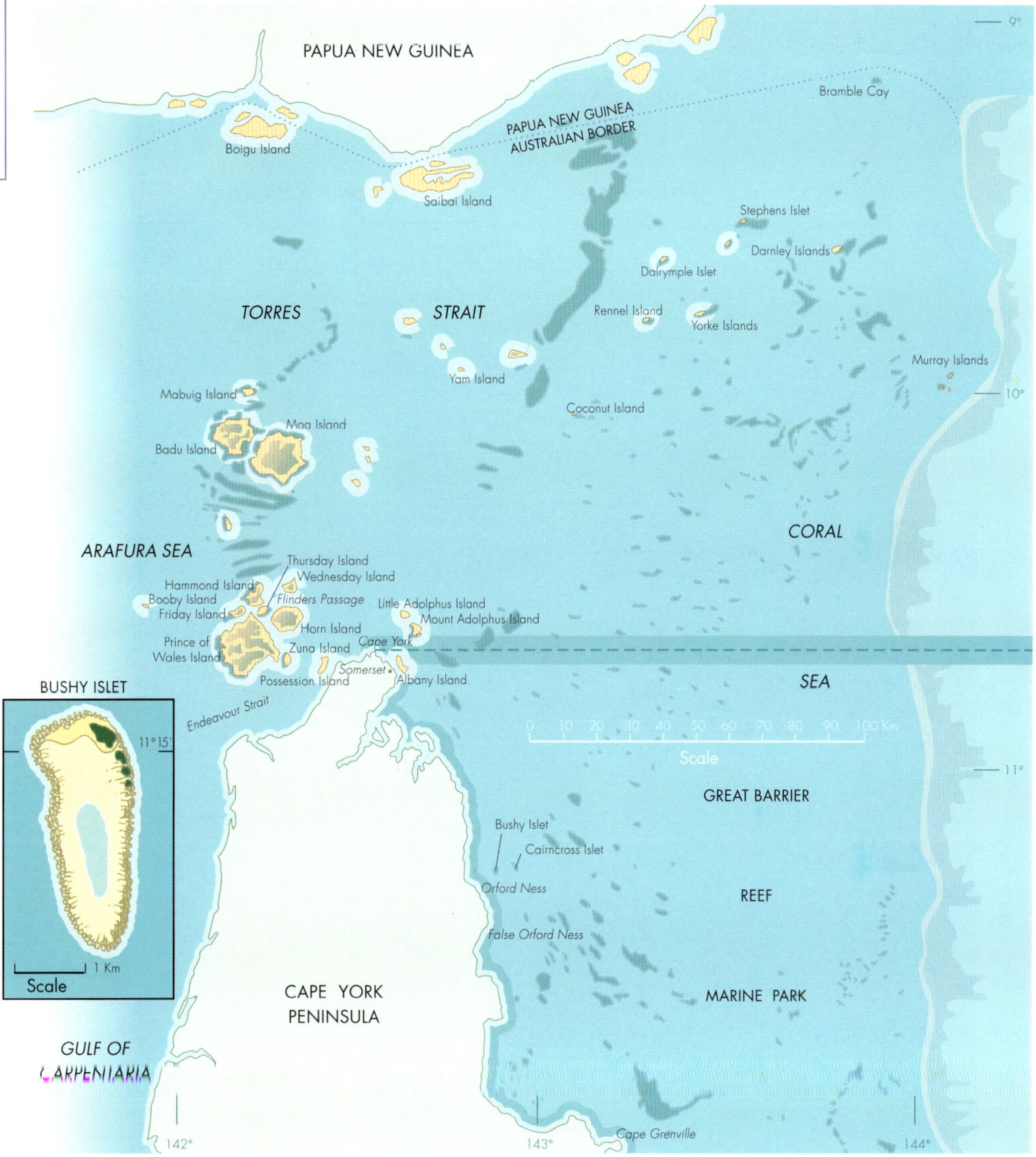

LOCATION

Albany (Pabaju) Island 10°44'S, Prince of Wales (Muralug) Island Bushy Islet 11°15'S, 142°52'E. Bushy Islet lies 7 km off the Cape York coast north-east of Orford Ness. Albany Island lies south-east of Cape York less than a kilometre off the mainland coast at Sommerset. Prince of Wales Island lies in a close group that includes Thursday (Waiben) and Horn (Nurupai) Islands, 20 km west of Cape York. Hundreds of islands stretch south across the Torres Strait from Bramble Cay (9° 9' S, 143° 52' E) for over 150 km.

AREA

Prince of Wales Island: 17,665 ha
Albany Island: 719 ha
Bushy Islet: 8 ha

CLIMATE

Tropical, monsoonal, strong trade winds

STATUS

Bushy Islet is vacant Crown land. The surrounding waters are part of the Great Barrier Reef Marine Park Far Northern Section. Prince of Wales Island is a mix of freehold and vacant Crown land. Albany Island is vacant Crown land with a leasehold portion for pearl-farming operations. Many of the islands and surrounding waters are the subject of pending claims under the Native Title Act from either Islander or Cape York Aboriginal claimants. The Torres Strait Treaty with Papua New Guinea provides for protection of the marine environment within the Torres Strait Protected Zone. The Great Barrier Reef Marine Park is managed by the Great Barrier Reef Marine Park Authority (GBRMPA) and the Qld Department of Environment. Albany Island is managed by the Qld Department of Natural Resources; other islands are managed by the Qld Goverment and local Islander councils.

ACCESS

Commercial air services to Thursday Island. Regular Torres Strait supply vessels call at Thursday and other inhabited islands.

FACILITIES

Full town facilities on Thursday Island, which is the commercial and administrative centre for the Torres Strait.

PAPUA NEW GUINEA

Bramble Cay

Boigu Island

PAPUA NEW GUINEA AUSTRALIAN BORDER

Saibai Island

Stephens Islet

Darnley Islands

TORRES   STRAIT

Dalrymple Islet

Rennel Island

Yorke Islands

Murray Islands

Yam Island

Mabuig Island

Moa Island

Coconut Island

CORAL

Badu Island

ARAFURA SEA

Thursday Island
Wednesday Island

Hammond Island
Booby Island
Friday Island

Flinders Passage

Little Adolphus Island
Mount Adolphus Island

SEA

Prince of Wales Island

Horn Island
Zuna Island

Cape York

Somerset

Endeavour Strait

Possession Island

Albany Island

Scale
0 10 20 30 40 50 60 70 80 90 100 Km

GREAT BARRIER

BUSHY ISLET

Bushy Islet

Cairncross Islet

Orford Ness

REEF

False Orford Ness

Scale   1 Km

CAPE YORK PENINSULA

MARINE PARK

GULF OF CARPENTARIA

142°   143°   Cape Grenville   144°

Cape York

Araucaria (Haslewood Island, Whitsunday Group)
Oil on canvas 122 x 107 cm

GEMS OF SAND, CORAL AND STONE

FLINDERS GROUP

HINCHINBROOK ISLAND

WHITSUNDAY ISLANDS

FRASER ISLAND

CAPRICORN BUNKER GROUP

Enclosed within the sweep of the Great Barrier Reef is a wondrous array of isles. Their origins are as disparate as the life they support: from inundated mountain ranges to constellations of coral and restless dunes born out of the currents. Turtles and terns, whales and wallabies, dugongs and dolphins—all call these islands and their environs home.

The Voyage of the Kangaroo

As Sydney and Hobart grew in the decades following 1788, so shipping movements increased between the fledgling colonies and the outside world. But the shortest routes to Asia and beyond were also the most dangerous, effectively barred by reefs everywhere to the north and north-east.

Early forays were not encouraging. In 1770, James Cook, the first to attempt a passage inside the Great Barrier Reef, almost met disaster among the hazards of what came to be known as the "inner route". HMS *Pandora* was wrecked in 1791 while negotiating the Torres Strait with a cargo of 14 of the *Bounty* mutineers, and in 1803, the *Cato* and the *Porpoise* struck reefs well east of Australia's tropical coast as they plied the "outer route". Numerous ships suffered the same fate elsewhere in the Coral Sea, across what was thought to be open ocean. Both inner and outer passages seemed practically impassable. To Matthew Flinders, aboard the *Porpoise* at the time of its wreck, voyaging north from Sydney required "nerves strong enough to thread the needle".[1]

The entire inner route is thought to have been first travelled in 1791 by a small group of runaway convicts in an open boat. But more than 20 years later, barely a ship of any size had dared run the gauntlet inside the reef. In April 1815, Lieutenant Charles Jeffreys left Sydney with a detachment of soldiers bound for Ceylon aboard HMS *Kangaroo*. Jeffreys intending to chance his luck with the outer route, but bad weather in the Coral Sea forced him to shadow the coast near Lady Elliot Island. *Kangaroo* was inexorably drawn into the maze. Working cautiously north and anchoring each night, Jeffreys reached the Endeavour River where Cook had careened his ship. Then, rather than seek open water as Cook had done, he continued north within the reef, eventually rounding Cape York in early June, after nearly two months of tense navigation. Jeffreys produced his own charts and added

considerably to knowledge of the inner route, which in time became the preferred passage. By the 1890s it was seen as the equivalent of a "relatively shallow and tranquil inland sea, which the largest steamers may traverse in safety".[2]

One hundred and fifty kilometres past the Endeavour River, Jeffreys found twin bays forming a great 85 kilometre-wide notch in the coastline. The broader he called Princess Charlotte Bay, and a group of craggy continental islands at its head Jeffreys named for his naval colleague, Matthew Flinders. In the following decades, most of the hydrographic expeditions exploring and charting Australia's northern waters called at this enormous bay bounded by Cape Melville. In addition to Flinders, the names ascribed to the islands and their surrounds read like a roll call of nineteenth-century hydrography; Phillip Parker King, Owen Stanley, Francis Blackwood, John Lort Stokes, Henry Denham, and the ships *Bathurst*, *Rattlesnake* and *Fly*.

Opposite: From near Howard Bluff at the northern end of Denham Island, looking across the entrance to Rattlesnake Channel to the Cape York mainland beyond.

Left: Fan Palms growing on the steep flanks of Stanley Island. Palms of the genus Livistona *are generally more fire tolerant than other species. They are found in many parts of Australia's wet-dry tropics and often grow on sandstone-derived or lateritic soils.*

Ships in the Stream

Heading for Princess Charlotte Bay in 1839, John Lort Stokes aboard HMS *Beagle* thought that "passing this point [Cape Melville] and Cape Flinders to be the most intricate part of the inner route".[3] The passage between coastline and reef has long since been charted and marked, but the dangers remain no less real. Some of today's hazards did not yet exist in 1815. To complicate matters, large ships, of a size and number unimaginable before 1850, now ply this passage. After Princess Charlotte Bay, they steer wide at speed past Cape Melville, change course across the shipping channel,

then steam south into a narrow stretch hemmed in by Warden Reef. The Great Barrier Reef is narrowest east of Cape Melville, with its outer edge just 20 kilometres from the coast.

Like Jeffreys, we sailed north by day and anchored at night. With some relief we negotiated our own approach, then passed the remarkable granite boulder pile of 400 metre-high Cape Melville — all through a channel free of looming ships. Sails aglow in the rich afternoon light, our yacht struck out on a broad reach across the 25 kilometre-wide expanse of Bathurst Bay. Clear of oncoming traffic, it was a relaxed but exhilarating passage towards Rattlesnake Channel and then on to our night's anchorage. The graceful profile of Flinders Island, its flanks rising steadily to 318 metre-high Flinders Peak, ushered us into the evening. Long after the short tropical twilight, we gingerly probed our way to the lee of Blackwood Island and settled for a peaceful night.

Remnants of a drowned extension of the rugged Bathurst Range, the four main islands in Flinders Group lie in a tight cluster commencing just over three kilometres across Rattlesnake Channel from the coast. Clack Island, a tiny outlier, is the furthest north, separated from Cape Flinders on Stanley Island by seven kilometres of open water. Geologically, the islands are part of the wider Cape York sandstone belt known as the Laura Basin, the basal layer of which outcrops on all the islands. Sediments up to 700 metres thick were laid down in a predominantly freshwater environment more than 140 million years ago during the Jurassic Period. The resulting sandstones, conglomerates and shales were then overlain by further sedimentary deposits in the succeeding Early Cretaceous Period. Topped by sparkling white quartz sandstone, these final deposits are rich in plant fossils, animal tracks, burrows and ripple marks.

By morning our seclusion was gone. Several prawn trawlers had returned from a night's fishing under floodlight on Bathurst and Princess Charlotte Bays. Their crews would spend the daylight hours recovering from the exertions of the night, before heading out the following evening to do it all again. Since the late 1960s the waters of the two bays have been a popular prawning ground worked by dozens of trawlers each year. The islands of the Flinders Group afford welcome shelter and secure anchorages. Years before, another fleet gathered in the protected waters west of Cape Melville.

Terrible Mahina

It was a calm, breathless evening inside the Great Barrier Reef. The monotonous tug of the south-east trade winds was strangely absent. From just inside Cape Melville to the Flinders Group, and west to the sweeping crescent of Princess Charlotte Bay, about 100 small boats rode at anchor. Their crews relaxed, enjoying the respite from strong winds and hard toil. Within a few hours most were dead.

By the 1890s, the burgeoning Torres Strait pearl industry was forced to work far from home, following the inevitable depletion of its own pearling grounds. Remote from their shore stations, luggers worked in fleets attended by schooners acting as motherships. Bathurst and Princess Charlotte Bays were oft-used anchorages. Flanked by the 500 metre-high Melville Range, they offered security from summer squalls and the incessant dry season

trade winds. By the accepted wisdom of the time the waters north of Cooktown were virtually cyclone-free.

One night in 1899, the barometer on a Bathurst Bay schooner dropped to the lowest level ever recorded in Australia, and within a few hours that belief had been blasted away for ever. Mahina, the terrible cyclone of Saturday 4 March, careered in from the Coral Sea and by 4.30 the following morning most of the fleet had been swamped and sunk, or dashed against islands and reefs. The eye passed and the gale backed. The remaining luggers were hurled onto the mainland shore by a surging tide. By Sunday afternoon over 300 Japanese, Malay, Filipino and Islander crew had perished. A handful of Europeans were also among the dead. About 100 Aborigines camped on the shore went to the aid of survivors, only to be swept away and drowned in the tidal surge.[4] Such was the fury that dolphins were flung more than ten metres up into the cliffs of Flinders and Denham Islands.[5]

Castles in the Sky

The shallow waters of Princess Charlotte Bay were streaked with white. It was late morning and the south-easterly was again blowing hard, so we decided to seek shelter within the Owen Channel. Flanked by the protective heights of Stanley and Flinders Islands, the channel is the most popular refuge within the group for the prawning fleet. After anchoring in the midst of about a dozen sleepy trawlers, I set out to explore Stanley Island.

In the lee of the Cape York Peninsula, Princess Charlotte Bay receives about 1,000 millimetres of rain annually, barely more than half that of Weipa, on the Gulf of Carpentaria coast. Relatively low rainfall and shallow sandy soils mean that the Flinders Group, like much of the Cape York mainland close by, manages to support only a mixture of semi-deciduous tropical woodland and shrub savanna.

I climbed up through deep spinifex and scrub dominated by straggly eucalypts, acacias and deciduous shrubs. The island's eastern flanks had looked readily negotiable from the narrow beach, now 50 metres or more below, but the vegetation barely restrained an unstable scree of sandstone rubble and loose boulders. Fighting my way to a band of sculpted grey sandstone rising ever more steeply into the cliffs of Castles Peak, I was enticed by a deep and inviting scallop of shade. Through a grove of stout Fan Palms (*Livistona muelleri*), rust-red inflorescences sprouting vigorously from their crowns, I clambered on, hoping to find a stretch of clear rock. But the way ahead was barred by cliffs.

Scratched, hot and plagued by biting Green Tree Ants (*Oecophylla smaragdina*), my attention was pleasantly diverted by a blur of pink bobbing about in an eddy of breeze. Protected by rumpled brows of stone, several sprays of the Cooktown Orchid (*Dendrobium bigibbum*) arched from a narrow fissure. Such perfect, orderly blooms were an exquisite contrast to the confusion of the surrounding scrub.

Outriggers

Looking down to the Owen Channel, I saw the prawn fleet slumbering on. Less than 100 years earlier, very different craft plied these waters. The traditional Aboriginal inhabitants of the Flinders Group and Bathurst Head nearby were the Walmbaria. They skilfully exploited the inshore resources of more than 90 kilometres of island coastline and intertidal reef flat. Their harvest included fish, molluscs, crustaceans, Dugong and turtles. With no suitable timber on the islands, the Walmbaria obtained canoes by trading stingray-barbed spears, throwing sticks and woven bags with coastal people from Princess Charlotte Bay. Superb seamen, they paddled as far south as the Endeavour River in their five metre-long single-outrigger canoes.

The Walmbaria occupied the islands for at least 2,500 years, with permanent camps on Stanley Island in enormous rock shelters on the north coast. Several metres of marine debris in the midden beds fronting the shelters speak of many generations' feasting from the sea.

With the arrival of Europeans came disintegration. As well as pearlers, trepangers and trochus shell fishermen also worked the region and disrupted traditional life. When ethnologists Herbert Hale and Norman Tindale visited the group in 1927, only 25 people were left on the islands and Bathurst Head. Most of their descendants now live in the Hope Vale Community near Cooktown. In 1994, freehold title to the Flinders Group (which had been a national park since 1939) was granted to its traditional owners. After the formulation of a management plan, the Flinders Group National Park is to be leased back to the Queensland Government. The park will be co-operatively managed.

Before retreating to the shore, I sidled around to the saddle between the two halves of Castles Peak. The westerly bluff, 220 imposing metres of quartz sandstone, broke free of the surrounding vegetation, sheer flanks towering over the island. Nagged by the trade winds from March to October and scoured by the summer monsoon, its sandstone fortifications were under near constant attack.

"… a remarkable cliffy lump …"

A spy hole had been blasted clean through a ridge-top blade of stone. It framed distant Clack Island, sitting on the edge of its own roughly oval-shaped reef like some morsel left on the edge of a plate.

Passing Cape Flinders, Stokes described Clack Island as "a remarkable cliffy lump". He also noted that the island was "interesting from the circumstance of Mr. Cunningham having found native drawings in its caves".[6] The botanist Allan Cunningham had accompanied Phillip Parker King on his survey voyages between 1818 and 1822. In June 1821, aboard the brig *Bathurst*, the pair made the last of three visits to the Flinders Group. Cunningham explored Clack Island looking for shells. However, he found much more. On the exposed south-east corner:

"…the weather had excavated several tiers of galleries; upon the roof and sides of which some curious drawings were observed … They represented tolerable figures of sharks, porpoises, turtles, lizards, trepang, starfish, clubs, canoes, water-gourds and some quadrupeds … Tracing the gallery round to windward, it brought me to a commodious cave … sufficiently large to shelter twenty natives. Many turtles' heads were placed on the shelfs … amply demonstrative of the luxurious and profuse mode of life these outcasts of society had at a period rather recently followed … this is the first specimen of Australian taste in the fine arts that we have detected…"[7]

Cunningham had stumbled upon one of many galleries of the Flinders Group and the adjacent mainland, the most significant body of Aboriginal rock art on the entire east coast. Clack Island was known as Ngurromo to the Walmbaria. It is a wild and dangerous place, difficult to reach and often impossible to land on. Of great importance in local mythology, access was restricted to initiated men concerned with the island's crowded galleries of sorcery art. Untended and exposed to the full blast of salt-laden air, sadly, Clack Island's spectacular art is rapidly fading.

In the waning afternoon, the narrow finger of Cape Flinders gestured a warning across a sea

Above: Animal motifs from the Ship Shelter, Stanley Island. The most extensive and varied rock art on the entire east coast is found in the Flinders Group. Motifs include fish, Dugong, birds, crabs and butterflies.

Below: "An example of the 'akaala' canoe from Flinders Island …" observed by Herbert Hale and Norman Tindale on their 1927 ethnographic expedition to North Queensland. Made of driftwood collected along the Bathurst Bay coast, such craft transported the Walmbaria and the mainland Mutumui people as far south as Cape Flattery and "to Cooktown in modern times" as well as to the Stewart River on the western shores of Princess Charlotte Bay. N.B. Tindale. Courtesy of South Australian Museum, Adelaide

Small wind-scoured cavities abound in the crumbly sandstone and conglomerate of the Flinders Group. Larger caverns provided shelter to the Walmbaria people for at least 2,500 years until as recently as the first decades of this century.

of burnished silver towards Clack Island. One hundred and seventy-five years before, the *Frederick* sailed from Hobart, only to be wrecked on the reefs fringing that cape. It was an early casualty of the "inner route", one of 650 vessels to be wrecked off the Queensland coast last century. Twenty-two crew were lost and five survivors rescued by the *Frederick*'s consort, the *Duke of Wellington*. Of shorter but still perilous journeys, no one will ever know how many Walmbaria, across unaccounted generations, failed to complete the voyage from the cape to Ngurromo and back.

The dipping sun cast a distant tanker in silhouette. Its great bulk would soon "thread the needle" past Cape Flinders and Cape Melville. From fragile canoes to luggers, trawlers and steel leviathans brimming with oil, these islands have seen an ocean of change. As I left the saddle for the descent to Owen Channel, I tried to imagine Mahina's fury as she hurtled in from the Coral Sea past Ngurromo. I also thought of the real possibility of such events recurring, catching just one of the thousand or more tankers that ply these waters each year. The inner route past these remote islands should never be regarded with complacency as just some "tranquil inland sea".

1   M. Flinders, *A Voyage to Terra Australis, Undertaken for the Purpose of Completing the Discovery of that Vast Country, in the Years 1801–1803*, G. and W. Nicol, London, 1814 (facsimile edn Lib. Bd of SA, 1966), vol. 2, p. 104.

2   W. Saville-Kent, *The Great Barrier Reef of Australia: Its Products and Potentialities*, London, 1893 (facsimile edn, Curry O'Neill, Melbourne, 1972), p. 3.

3   J. L. Stokes, *Discoveries in Australia; With an Account of the Coasts and Rivers Explored During the Voyage of H.M.S. Beagle in the Years 1837–43*, T. and W. Boone, London, 1846 (facsimile edn Lib. Bd of SA, 1969), vol. 1, p. 352.

4   As a gesture of appreciation for assisting the survivors of Mahina the Queensland Government gave the surviving Aborigines "100 red shirts, 100 turkey-red dresses, eight dozen tomahawks, one ton of flour, two gross of pipes, 173 1/2 pounds of tobacco and 200 knives". The cyclone had left them on the brink of famine: their hunting and foraging grounds decimated and their canoes in pieces. H. Holthouse, *Cyclone*, Rigby, Adelaide, 1971, p. 14.

5   The Mahina tragedy was a great blow to the Cape York pearling industry but did not mark its end. Two hundred and fifty luggers were still working out of Thursday Island at the turn of the century. As a maritime disaster, the destruction of the pearling fleet is exceeded only by the sinking of HMAS *Sydney* in World War II and the loss of the immigrant ship *Cataraqui* in 1845 as Australia's worst maritime disaster.

6   Stokes, op. cit., p. 352.

7   Allan Cunningham's manuscript quoted by P. P. King, *Narrative of a Survey of the Intertropical and Western Coasts of Australia, Performed Between the Years 1818 and 1822*, John Murray, London, 1827 (facsimile edn Libraries Bd of SA, 1969), vol. 2, pp. 25–7.

Cunningham's find was only the third European record of Aboriginal rock art in Australia. Engravings had previously been noted by Arthur Phillip at Port Jackson in 1788, and in 1803 Matthew Flinders gave the first descriptions of rock paintings on Chasm Island near Groote Eylandt in the Gulf of Carpentaria.

FLINDERS GROUP

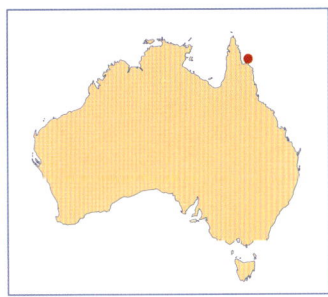

LOCATION
14°10'S, 144°18'E.
Lies between Bathurst and Princess Charlotte Bays, commencing 3 km off Bathurst Head and 150 km north of the Endeavour River. The closest town is Cooktown.

AREA
Flinders Island: 1,480 ha
Stanley Island: 870 ha
Denham Island: 485 ha
Blackwood Island: 178 ha

CLIMATE
Tropical, monsoonal—moderate rainfall due to the rainshadow of Cape York.

STATUS
Flinders Group National Park. Surrounding waters are part of the Great Barrier Reef Marine Park, Far Northern Section. Freehold title to the islands will be granted to Aboriginal claimants from the Hopevale Community after the completion of a plan of management, due in 1997. Currently managed by the Department of Environment, Queensland. Joint management with the Aboriginal claimants is planned. Surrounding waters are administered by the Great Barrier Reef Marine Park Authority (GBRMPA).

ACCESS
Cape York and Torres Strait supply vessels sometimes stop at the Flinders Group as do some Great Barrier Reef tourist cruise vessels. Good cruising yacht anchorages. Camping permitted on all but Denham Island—contact Department of Environment for details.

FACILITIES
Boardwalks and interpretive information at major rock art galleries.

CORBETT REEF

Clack Island
Clack Reef

King Island

CORAL SEA

GREAT

BARRIER

REEF

14° 00'

144° 15'

144° 30'

Pipon Island
Lighthouse

Cape Flinders   FLINDERS
Castles Peak 220
Stanley Island   GROUP   Cape Melville
Channel

Flinders Island
+ Flinders Peak 318

MELVILLE RANGE

14° 15'

Fly Channel

Blackwood Island   Maclean Island

Rattlesnake Channel   Denham Island   BATHURST

BAY

South Warden Reef

Bathurst Head

NINIAN

PRINCESS

BATHURST RANGE

BAY

CHARLOTTE

Barrow Island
Lighthouse

BAY

CAPE YORK PENINSULA

Barrow Point

NORMANBY RIVER

0   5   10   15   20 Km

Scale

HINCHINBROOK ISLAND

Summit Panorama

Mount Bowen is the high point of a granite massif at the heart of Hinchinbrook Island. The peak is crowned with grass trees, wind-stunted banksias and granite tors blotched with lichens. On a clear day the views from this summit extend to the island's farthest shores. The northern flanks of the mountain plunge to low-lying mangroves, fretted with tidal creeks. This expanse shelters behind an arc of dunes and beach that tethers the distant outcrops of Cape Sandwich and Cape Richards to the main bulk of the island. A chain of lower peaks stretches for 24 kilometres along the north-western arm of the island to Hecate Point. To the east, steep spurs and creeks descend to rocky headlands and crescent-shaped beaches. Looking west and south, there are more granite ramparts and deep valleys cloaked in dense rainforest. Beyond lies Hinchinbrook Channel, fringed with mangroves. This passage, in places less than a kilometre wide, is all that separates the world's largest island national park from the mainland.

Warrawilla In The Mist

When conditions are favourable this vista is one of the most spectacular offered by any Australian island. But clear skies are the exception at the summit. The massif is almost invariably wreathed in cloud, as we discovered one warm November afternoon. After a long day climbing the rock-strewn Warrawilla Creek from the beach at Little Ramsay Bay, the upper 300 metres of this 1,121-metre peak was mantled in dripping mist. The headwaters of the creek is a near-vertical scree of immense boulders. Even when the cloud cover eased slightly, the view was, in turn, obscured by an enclosing canopy of rainforest.

The late afternoon light filtered through the foliage of tall oaks, Golden Pendas (*Xanthostemon chrysanthus*) and Corkwoods (*Melicope elleryana*). A tangle of vegetation crowded the banks of the creek. Ferns and palms flourished, including the dreaded Lawyer Vine (*Calamus* sp.). These palms feature attractive spreading fronds and great loops of tentacle-like stems decorated with savage barbs, ready to hook into passing flesh and clothing. In fading light the only practical option was to follow the watercourse. The sodden creek stones and fallen logs were matted with lichens, mosses and algae, making progress a precarious balancing act.

It was too late to push on to the saddle leading to the summit of Mount Bowen. Before trying to make camp I was determined to climb out onto the face of the mountain. Ostensibly I wanted to try to fix our position but mostly I needed to dispel the feeling of entrapment created by the entwining undergrowth. I broke out of the creek by crawling between a thicket of creepers and onto a steep slope of loose rock and forest debris.

I followed a tributary of the main creek that cascaded over slabs to the right of a tall buttress of dark rock. The only way up was to climb on the narrow strip of water-scoured rock using the sturdy grasses alongside as handholds, with spray cannoning down onto me. Eventually I clawed onto a grassy ledge that cut across the buttress at half-height, while plumes of heavy cloud swirled above and below. Occasionally there were glimpses of rock faces streaming with water on the northern side of the creek. For a few short seconds the mist parted and I gazed all the way down the valley to the brilliant white sands of Little Ramsay Beach and a clear, blue horizon. Then the curtain of cloud once again wrapped tight around the mountain.

Back in the main creek we sheltered for the night in a horizontal fissure that bisected an enormous perched boulder. At nightfall bats began their patrols under the still dripping foweest canopy. On our narrow rock shelf I cooked a meal in the company of an inquisitive Fawn-footed Melomys (*Melomys cervinipes*), a small native rodent. The damp air hummed

The view north from the summit of Mount Bowen takes in the vast mangrove communities of Missionary Bay, nearby Goold Island and the mainland beyond.
Lincoln Hall

with insect life and through the night bird cries echoed off the walls above. As the darkness became absolute it seemed as though we had taken refuge in the jaws of a fantastic creature standing at the gates of creation.

Massif Myths

The peaks of Hinchinbrook have always loomed large in the human history of the area. Of the Aboriginal groups who have traditionally inhabited the region, those who made the island their home are known as the Bandyin people, a title with direct links to the mountains of the island. Their mythological naming of the landscape reflected a belief in the power of ancestral creation spirits that resided in the mountains. One of the better known of these is Gunandali (Mount Diamantina), the distinctive summit in the south of the island capped with an enormous cleft block.

The first Europeans to look upon these shores were similarly struck by the island's topography. In the early hours of 9 June 1770 HMS *Endeavour* stood off the coast. With the aid of a light moon Captain James Cook gazed upon the skyline between Hillock Point and Cape Sandwich and wrote in his journal "the

Land between them is very high, and of a craggy, barren surface."[1] Nearly 50 years later, Captain Phillip Parker King surveyed this shore at closer quarters and noted that the terrain had "a singularly grand and imposing appearance. It rises nearly perpendicularly from the lower wooded hills at its base, and is as abrupt on its land side as on that which faces the sea."[2]

King, in command of HMS *Mermaid*, explored Goold and Brook islands in Rockingham Bay. He was the first European to suspect that the opening to the south separated the massif from the mainland. This was confirmed by the survey of HMS *Fly* in 1843. In his narrative of the voyage the naturalist J. Beete Jukes wrote of Hinchinbrook: "This is the most singularly broken mass of hills I ever saw, covered with rugged knolls, and sharp inaccessible pinnacles, and furrowed by deep and precipitous gullies and ravines."[3]

Uplifts And Downpours

This formidable topography had its origins in major uplifts and folding of the earth's crust some 260 million years ago. At this time the granite that distinguishes Hinchinbrook's lofty heights intruded into older silicic-volcanic rocks, of the kind still evident in the range along the north-western arm of the island. Erosion of the older rock layer subsequently left the more resistant granite on peaks like Mount Diamantina and Mount Bowen.

This distinctive rock type is not confined to the summits, but is widespread on ridges forming monolithic prows and spires, giving the island an almost alpine appearance. The processes of weathering and erosion have also exposed the granite in talus slopes, broad stone pavements, waterfalls and ravines. Streams draining the massif are typically jumbled with gigantic blocks and rounded boulders, sculpted and bludgeoned by the run-off from seasonal storms.

Most of the island's rain falls in torrential downpours during the months between December and April. The region around Hinchinbrook stands on the threshold of the wet tropics. Within this broad categorisation there is considerable local variation in climate. Tully, 20 kilometres to the north of the island, is renowned as Australia's wettest town, with an average annual rainfall of 4,206 millimetres. Meanwhile Cardwell, on the mainland immediately adjacent to the island, receives a relatively modest 2,143 millimetres a year. No detailed records exist for the island itself, but estimates for Mount Bowen suggest an annual quota exceeding 3,500 millimetres.

Off The Shoulder

In the midst of the forest, high in the catchment for Warrawilla Creek, such a statistic seems perfectly plausible. After a night of fitful sleep I woke in our granite penthouse to the sound of rain dropping through the forest. With the probability of more downpours and persistent, soupy cloud, we retreated from the mountain. Within a few hundred metres we re-entered a world of dazzling sunlight and bleached creek stones.

The transition from the luxuriant growth of the musty forest to the heath and scrub plants on the mountain's exposed shoulders was startling. Leaving the creek, we made slow progress through a dense cover of Grass Trees (*Xanthorrhoea* sp.), She-oaks (*Casuarina* sp.),

Banksias (*Banksia integrifolia*) and Tea-trees (*Leptospermum* sp.) After so much time in the open corridor of the creek, I suddenly felt as though I were being jostled and prodded by a hostile mob. Much of the undergrowth was abrasive or armed with spikes, and the thicket of low branches was often impenetrable.

Yet, for all the obstacles, this scrub and heath landscape bristles with interest. Clearings formed by granite slabs and pavements are colonised by carpets of the Resurrection Plant (*Boyra septentrionalis*), a diminutive shrub which dries to shades of vibrant orange. Other shrubs and wildflowers like Boronia (*Boronia rosmarinifolia*), Fringed Lily (*Thysanotus tuberosus*) and Wild Iris (*Patersonia glabrata*) give the bush vivid splashes of pink and purple. At ground level delicate grasses and sedges sprout from the dark, sandy soil, which in places is studded with nuggets of white quartz. While Rainbow Lorikeets (*Trichoglossus haematodus*) and Yellow-spotted Honeyeaters (*Meliphaga notata*) feast on nectar, tiny Scrubwrens (*Sericornis* sp.) flit through the thickets.

Then, in a patch of scrub recently cleared by fire, unmistakable signs of new growth marked the charred and desolate landscape. Grass Trees sported tufts of fresh greenery, young grasses were emerging from the ash beds and tender shoots stood out on the blackened trunks of small eucalypts.

Burning For Banksias

A century ago such burnt areas would have been commonplace on the island. The Bandyin people, like many other groups on the mainland, routinely set fire to areas of scrub to assist them in hunting animals, gather edible roots and manage the landscape. In the decades since such 'firestick farming' ceased, the diversity of Hinchinbrook's plant communities has been put at risk. Those species dependent upon fire for seed germination and dispersal, and others that benefit from the clearing of leaf litter, have suffered from the lack of regular burning.

In recent years the Queensland National Parks and Wildlife Service, in conjunction with researchers from James Cook University, has

Regrowth after controlled burning, designed to help preserve the mosaic of plant communities on Hinchinbrook Island.

Fruit of the Blue Quandong (Elaeocarpus grandis) *on the rainforest floor.*

Creeks converging at Zoe Bay help sustain the rainforest.

carried out prescribed burns, aimed at restoring the mosaic of plant distribution on the island. Two key beneficiaries of this program are *Comesperma* sp., a plant endemic to the Hinchinbrook heaths, and Dallachy's Banksia (*Banksia plagiocarpa*). This species, unrecognised until 1979, is found only on Hinchinbrook and parts of the adjacent mainland.

On the lower slopes of the island, the scrubs and heaths give way to woodlands and tall open forest. Here deeper soils support a mix of species including Bloodwood (*Eucalyptus intermedia*), Turpentine (*Syncarpia glomulifera*), Moreton Bay Ash (*Eucalyptus tessellaris*) and wattles (*Acacia* sp.). Moister areas, such as soaks and freshwater swamps, are dominated by paperbark (*Melaleuca* sp.) woodlands, often with an understorey of *Pandanus*. Fire has also traditionally helped maintain many of these communities, particularly as a border defence against encroaching rainforest.

Forest Domains

To walk on Hinchinbrook is to be constantly confronted by abrupt changes in vegetation and habitat. Our descent from the mountain took us through dense She-oak thickets and open sclerophyll forest; from rocky ridges of Tea-tree scrub to stands of huge Paperbarks. Most remarkable of all were the sudden transitions to tall rainforest, such as the expanse dominating the hinterland of Zoe Bay.

From the dappled brightness of sere eucalypt woodland we plunged into a lush, shadowy realm. Towering canopy species like Blue Quandong (*Elaeocarpus grandis*), Red Mahogany (*Eucalyptus resinifera*) and Milky Pine (*Alstonia scholaris*) enclose the rainforest. These trees, with some buttressed specimens rising more than 30 metres, are festooned with vines, tendril climbers and epiphytic ferns. Others have fallen prey to the gruesome embrace of Strangler Figs (*Ficus watkinsiana*). Below, the cool, lime-green light of the interior supports a profusion of palms, including the Fan Palm (*Licuala ramsayi*) with its elegant umbrellas of radial fronds.

The richness of the forest, the feeling of completeness and diversity is inescapable. Just when everything appears to merge into a palette of greens and rich browns, one's gaze fixes on the startling cream and orange stripes of a fungus, or a scattering of gentian-blue Quandong fruit on the forest floor. Similarly the stillness and silence is broken by the cracking call of a Victoria's Riflebird (*Ptiloris victoriae*), or the sound of a skink ruffling through the matt of leaf litter at one's feet. Everywhere there is a humbling sense of scale, abundance and archaic rhythms.

These habitats are important refuges and foraging grounds for an array of gaudy and vocal birdlife. This ranges from the large Australian King Parrot (*Alisterus scapularis*) with

its brilliant red plumage, to the small but colourful Noisy Pitta (*Pitta versicolor*), and the Spotted Catbird (*Ailuroedus melanotis*) with its distinctive meow-like call. Several species of fruit doves and pigeons also feed in the forest. These include large flocks of Torres Strait or Nutmeg Pigeons (*Ducula spilorrhoa*) that visit Hinchinbrook from their nesting grounds on the neighbouring Brook Islands. Fallen fruit is scavenged by inhabitants of the forest floor, among them the Southern Cassowary (*Casuarius casuarius*).

Teeming Tides

The rainforest at Zoe Bay butts directly onto the beach and the mangrove-fringed inlets, where mountain creeks meet the sea. Near the coast, the intermittent noises of the forest are underscored by the sound of waves rushing onto the shore. After the cool shade of the forest, the sunlight glinting off the white sands of Zoe Beach is stunning. By the time we reached the northern end of the beach it was low tide and the mangroves that surround the estuary of North Zoe Creek were teeming with life. Exposed by the receding waters, the oozing mud bubbled and popped. Great Mudskippers (*Periophthalmadon* sp.) darted over the surface, while Fiddler Crabs (*Uca coarctata*) fossicked for food, stopping only to raise their large orange claws in comic salute. Though only recently revealed by the tide, the grey-blue mud was already dotted with freshly fallen leaves and bright scarlet flowers of the Large-leafed Mangrove (*Bruguiera gymnorrhiza*).

While this mangrove pocket is the largest on the east coast, it is easily dwarfed by the dense tracts that lie in the protected waters off the island's western flanks. Mangroves constitute one of Hinchinbrook Island's most outstanding natural assets, with some 31 species represented. The vast areas colonised in Missionary Bay and Hinchinbrook Channel form one of the largest remaining stands of mangroves in Australia.

This environment had its origins as sea levels rose at the end of the last ice age. The island became separated from the mainland with the inundation of Hinchinbrook Channel some

10,000 years ago. Over the millennia, fluvial deposits from the Herbert River have progressively formed sediment beds and islands along the channel. These nutrient-rich deposits lying close to the surface are exploited by the varied mangrove fauna. Elaborate root structures serve to stabilise the silt and mud. At the same time these roots supply essential nutrients, air and water, as well as helping to control salt levels. The Herbert River delta is also the most southerly known habitat of the Nypa Palm (*Nypa fruticans*), a salt-tolerant species with fronds as long as 10 metres.

The dense expanses of mangrove forest are broken only by a network of narrow tidal

In its headlong descent from Mt Diamantina, South Zoe Creek has created a succession of spectacular waterfalls and rock pools.

107

View of Goold Island and Mt Hinchinbrook, by Phillip Parker King. The Mermaid *lies in the middle distance. Courtesy of Mitchell Library, State Library of N.S.W.*

and was further enlarged in 1960 with the resumption of the outstanding leasehold land.

While mangroves dominate the western side of the island, the focus for most visitors to the island is the east coast, which is characterised by an indented shoreline with a succession of beaches and headlands backed by rampant forest. In the years since acquiring park status, the island has been left essentially undeveloped, with the exception of a small, low-impact tourist resort at Cape Richards.[4] Aside from visiting boats, the majority of travellers to the island are bushwalkers, most of whom trek the 32-kilometre Thorsborne Trail. This marked track is maintained by the Queensland National Parks and Wildlife Service and takes in the natural splendours of the east coast. By controlling numbers and encouraging minimum-impact camping practices, this trail is designed to offer an experience in keeping with the wilderness character of Hinchinbrook.

On our last day on Hinchinbrook we joined this trail for the journey north. The fellow travellers we encountered seemed young, fit and eager to witness the natural splendours of the island. It was an enthusiasm not hard to share as we padded along the scalloped sands of Little Ramsay Beach. A regimental line of casuarinas faced the beach. Flecks of grey dotted the mineral sands as the waves pushed tongues of foam onto the shore. We scrambled around headlands of pink granite to Nina Bay and then on to reach the sweeping eight-kilometre frontage of Ramsay Bay. I dropped my pack and plunged into the warm waters of the Coral Sea. Floating on my back I looked up at the mountains, with the spire of Nina Peak in the foreground. Beyond, stood the jagged summit skyline of Mt Bowen, tantalisingly cloud-free.

creeks. These form an essential habitat for an astonishing variety of marine life. Several species of crustaceans, molluscs and other invertebrates consume plant matter from the mangroves and are, in turn, a food source for fish and birds, including Great-billed Herons (*Ardea sumatrana*) and four species of Kingfisher (*Alcedo* sp. and *Halcyon* sp.). At the apex of this food chain are Estuarine or Saltwater Crocodiles (*Crocodylus porosus*), which roam the tidal creeks and bask on mud banks and nearby beaches during the ebb tide.

On The Trail

For the Bandyin people the island's shores also provided a rich and varied source of food. Though it is 100 years since they were dispossessed, the evidence of their successful island existence endures in the form of numerous middens and a series of fish traps made from rock enclosures at Scraggy Point. European attempts to settle on land nearby and on the shores of Missionary Bay were short-lived. The island was first declared a national park in 1932

1   J.C. Beaglehole, *The Journals of Captain James Cook on His Voyages of Discovery*, vol. 1, Hakluyt Society, Cambrige, 1955.

2   P.P. King, *Narrative of a survey of the intertropical and western coasts of Australia performed between the years 1818 and 1822*, John Murray, London, 1827, vol. 1, p. 109.

3   J.B. Jukes, *Narrative of the surveying voyage of HMS* Fly

commanded by Capt. F.P. Blackwood, T. & W. Boone, London, 1847, pp. 90-93.

4   At the time of writing, controversial entrepreneur Keith Williams is in the process of building a 2,000 bed resort and 234-berth marina at Oyster Point on the mainland near the southern entrance to Hinchinbrook Channel.

HINCHINBROOK ISLAND

LOCATION

18° 22'S, 146° 15'E. 100 km north of Townsville. The island spans the coastline between the townships of Lucinda and Cardwell

AREA

39,300 hectares

CLIMATE

Humid tropical with average annual rainfall exceeding 3500 mm on parts of the island.

STATUS

Hinchinbrook Island is a National Park managed by the Queensland Department of Environment. The waters surrounding the island are part of the Great Barrier Reef Marine Park.

ACCESS

Either by private vessel or charter boats from nearby Cardwell and Lucinda.

FACILITIES

The Thorsborne Trail is a 32 km marked bushwalkers trail along the island's eastern coastline from George Point to the southern end of Ramsey Bay. Bush campsites and short marked walking tracks at Scraggy Point and Macushla Bay. Permits are required to camp and for extended bushwalking on the island. A small, secluded commercial resort operates at Cape Richards, on the northern extremity of the island.

---

0   5   10 Km
Scale

CORAL SEA

Goold Island

Brook Islands

ROCKINGHAM

BAY

Cape Richards

Cape Sandwich

Hecate
Point

Missionary Bay

CARDWELL

The Haven

RAMSAY BAY

Hinchinbrook Channel

Nina Peak

Mt Bowen (1121)

Warrawilla Creek

Hinchinbrook Island

Zoe Bay

Mt Diamantina (955)

Mulligan Bay

George Point

LUCINDA

— 18° 10'

— 18° 20'

— 18° 30'

146° 10'   146° 20'

WHITSUNDAY ISLANDS

A *Whirlybird's Eye View*

Mark Luyten is one of 10 rangers responsible for approximately 45,000 square kilometres of the Great Barrier Reef Marine Park. In one sweep of water, roughly spanning the degree of latitude between the towns of Bowen and Mackay, are 160 islands or islets, 208 square kilometres of national park and 10 major tourist resorts. Each year this region is besieged by 350,00 visitors eager to claim their share of paradise. To keep a watchful eye on the park's natural assets the rangers conduct regular patrols by sea and air. On a warm October morning we joined Mark for a helicopter survey of the Cumberland Islands, a region better known simply as the Whitsundays, the largest island chain in eastern Australia.

Enclosed by wooded hills, the modest airstrip inland from Shute Harbour gave no clue as to what lay over the ridge. However, within a minute of take-off, the vista of islands scattered across this stretch of the Coral Sea was spread before us. We left the coast behind and flew south-east, passing Hamilton Island, an arresting vision for anyone who thrills to the sight of an island humbled by an airstrip, marina and high-rise towers. Neighbouring Whitsunday Island was a refreshing contrast: a mass of lofty green ridges and a coastline fretted by inlets and curving bays.

We whirred on, crossing a long stretch of open water to Haslewood Island. As we swooped low over reefs hugging the island's coastline, the water graded from shades of deep ultramarine to soft powder blues. White sails dotted the waters south to the Lindeman Group.

Veering back west to Whitsunday Island, we followed the swirling sands of Hill Inlet up onto the island's rocky spine. Tall forest appeared to rise to meet us, with palms breaching the canopy, seemingly close enough to touch. Then from Whitsunday Cairn, at the northern point of the island, we skimmed back to the east to Border Island, before making the clear run north-east to Hook Island. There we lobbed gently onto the shore at Mackerel Bay and tumbled out onto a blinding white beach. Our rollercoaster ride was over.

Opposite: Looking south-west along shores of Haslewood Island to the distinctive profile of Pentecost Island on the horizon.

Left: Two of the many species of Giant Clam found on the fringing reefs in the Whitsundays.
Tony Fontes

Caught On Hook

Hook Island's outstanding features are two parallel inlets, Nara and Macona, which reach deep into the heart of the island. Deep water and encircling ridges make these fiord-like bays among the busiest yacht anchorages in the area. To the east of the inlets, at the gateway to Hook Passage, is a small resort and underwater observatory. Meanwhile, on the opposite side of the island a steady flotilla of boats make their way north to the opulent resort at Hayman Island and popular anchorages at the top end of Hook Island.

While tourism gnaws at the edges of the Whitsunday Islands, the wooded valleys and steep ridges that dominate their interiors remain largely unscathed. After a night camped on the sandy rim of Mackerel Bay I climbed a ravine behind the beach and headed inland. A

forest of ferns and paperbarks enclosed the shadowy, boulder-choked gully. At the top I was greeted by a cooling breeze and a daunting barricade of sheoaks and grass trees. At the summit I slumped onto a dark terrace of basalt spattered with green and white lichen. In the foreground, across a timbered defile, stood Hook Peak, at 459 metres the highest summit in the Whitsundays. Sharp crests and lush, forested valleys extended southwards, merging with those of Whitsunday Island to form a 30-kilometre expanse of rugged terrain. In the milky haze of early afternoon the surrounding sea looked like a bank of infill cloud around a veritable cordillera of peaks.

Around 18,000 years ago this hallucination was a topographical reality. At the time of the last great ice age these islands were the caps of a mountain range straddling an extensive coastal plain. The peaks had their beginnings

Stately Hoop Pines guard the rocky fringes of the islands in the Whitsunday Group, as seen here on a ridge overlooking Mackerel Bay on Hook Island's north-east coast.

Superbly skilled at woodland camouflage, the Tawny Frogmouth (Podargus strigoides), seen here on Long Island, enjoys the night life and a diet of insects, small lizards and spiders.

approximately 65 million years earlier, following the break-up of Gondwana. As Australia drifted northwards, the eastern edge of the liberated continent passed over 'hot spots' in the earth's mantle. Volcanoes formed at breaches in the continental crust, releasing massive volumes of molten lava which cooled to become basalt. Meanwhile, below the surface, granite formed. The weathered remnants of these eruptions and implacements were eventually marooned by rising seas 6,000 years ago, creating the present-day island congregation.

Leaving the summit's basalt cap, I picked my way down the steepening slopes. In a few strides I entered a lush vine scrub, the kind of dank locale not extolled in any 'sunlover' holiday brochure. Under the canopy of trees and palms the air hung still and musty. I slithered down moss-coated boulders. In such forests it is forever autumn. Apart from the occasional bird rustling on a high branch, the only movement was the steady drift of leaves.

From the base of a small waterfall I stumbled out of the bush and onto the gravel of the main watercourse draining from Hook Peak. It was the end of the 'dry' season and the creek stood empty. But come summer such creeks can be transformed into surging torrents. While the region is renowned for its equable climate, with mostly light winds and gentle seas, it falls within the path of cyclones. At 2 a.m. on 17 January

1970 the eye of cyclone Ada passed over Hook Island. Huge seas and savage winds devastated resorts and towns. Fourteen lives were lost and hundreds were left homeless. An anemometer on Hayman Island recorded windspeeds of 160 knots before being blown away.[1]

Looking upstream there was no hint of such forces of destruction, just a bank of high ridges dotted by outcrops of basalt, some forming caves and long, dark overhangs. For at least 10,000 years natural shelters like these were used by members of the Gia and Ngaro clans. It is thought that more than 300 Aboriginal people inhabited these islands and their surrounds.

Evidence of this maritime life is found in shell middens, stone fish traps and quarries, such as those on South Molle Island, where stone was excavated for tool making. At Nara Inlet are also some rare examples of island rock art: enigmatic barred circles and zigzag shapes drawn in red and white ochre—paintings whose full meaning has vanished with their creators.[2]

Fire Sticks And Haslewood

After three days at Mackerel Bay we hitched a ride aboard a Marine Parks service boat skippered by Paul Harrison, another of the park rangers. Our first stop was outlying Border Island where another half-dozen yachts bobbed at anchor in Cateran Bay. From the boat we gazed up at the eastern flanks of the bay. The slopes leading to the summit of Mosstrooper Peak were singed in a patchwork of brown and black, the result of a recent prescribed burn.

A feature of Aboriginal life in the Whitsundays, as elsewhere, was the practise of burning patches of scrub. In the 200 years since European occupation the mix of vegetation across the gamut of island habitats has steadily altered, partly due to the lack of intermittent burning. Rainforest has encroached on areas of open woodland, dominated by the Poplar Gum (*Eucalyptus platyphylla*). Elsewhere, many grassland species have been out-competed by invasive weeds. In response to this loss of habitat diversity, controlled burns are being carried out on islands where ecosystems are imperilled.

Some of the best preserved vegetation communities in the Whitsundays are found on Haslewood Island. Unlike Hook Island's soaring ridges and valleys, the topography on Haslewood is more subdued. Its undulating hillsides are covered by a mosaic of woodland and tall tropical grassland. Such associations have all but disappeared on the adjacent mainland, and islands like Haslewood are believed to represent "the best chance for preserving native tropical grasslands along coastal Queensland."[3]

We landed at Windy Bay on a sweep of dazzling white beach, which forms Haslewood's north-western point. That night the serenity of our camp was disturbed by the sound of something hefty crashing through the vine scrub. The next morning I discovered the unmistakable tracks and wallows belonging to

Haslewood's population of feral pigs. They kept out of sight, but another exotic species, Lantana (*Lantana camara*) was unavoidable. It formed an impenetrable thicket in shallow gullies.

I crossed a broad undulating plateau dominated by casuarinas and eucalypts. Among these dominant trees was the occasional Pandanus, its outstretched limbs making it appear like some grounded prehistoric bird. Elsewhere were pockets of bright green palms and stone outcrops cradling fig trees. I emerged from this woodland onto long grassy slopes that dropped to the shores of White Bay. A steady south-easterly breeze raced across the water, tossing the grass into flowing manes. Hundreds of Grass Trees (*Xanthorrhoea arborea*) were assembled across the slope like mute sentries, their long, upright flower stems swaying in the wind.

Fallen fronds of Pandanus tectorius *on coral rubble.*

The exposed granite battlements of Seaforth Island in the Lindeman Group.
Rob Jung

Looking south from Shaw Peak on Shaw Island in the Lindeman Group.
Rob Jung

Rites Of Passage

From a small rock outcrop I looked south across an empty stretch of water to where an isolated volcanic cone stands proud on the horizon. Of all the islands HMS *Endeavour* passed by on 3 June 1770 it was this singular formation that James Cook described as being "more remarkable than the rest." The day of its sighting being the seventh Sunday after Easter he called it Pentecost Island. And the slender strait it helped guard was named Whitsunday Passage.

In the years that followed Cook's 'discovery', many other vessels steered a course through this deep passage, including merchant ships and boats bearing fugitive convicts north in a bold bid for freedom. In 1815 Charles Jeffreys, aboard HMS *Kangaroo*, took shelter in the lee of Long Island during the earliest documented voyage along the inner route of the Great Barrier Reef. Four years later, HMS *Mermaid* anchored off neighbouring Pine Island and Phillip Parker King became the first European to 'officially' make landfall on the Cumberland Islands.

My rocky vantage point was shaded by an impressive stand of Hoop Pines (*Araucaria cunninghamii*), the dominant tree of these luxuriant shores. When King briefly revisited the Cumberland Islands in July 1820 he was accompanied by the botanist Alan Cunningham. Both men were intrigued by the stately pines. To King they were a potential source of ships spars; for his companion, this distinctive tree was yet another discovery for science. For the Gia and Ngaro people these reports of resplendent forests were the first in a series of fateful disclosures about their island domain.

By the 1860s Whitsunday Passage was a busy thoroughfare for ships supplying newly established coastal ports such as Bowen and Mackay. As well as coastal traders there were bêche-de-mer luggers and ships calling to take timber. Some of the encounters between the Aborigines and these interlopers were peaceful, but many ended in chaos and bloodshed. In two decades a spiral of violence and reprisal saw the disintegration of "a dynamic and wonderfully adapted society."[4] Looking out over waters once plied by canoes, no tangible sign of those grim years remained, just the pines sighing in the wind and ranks of Grass Trees, their spears raised in defiance.

On The Fringe

By mid-afternoon the tide had withdrawn from White Bay. I walked out across an immense fringing reef. Bathed in sunlight, the discs and domes of coral shone in luminous shades of yellow, purple, orange and magenta, as though lit by neon.

While I stepped tentatively among the formations, crabs scattered for cover and Giant Clams (*Tridacna* spp.) squirted shut, their fleshy mouths forming rumpled grins like cartoon villains. On the landward margins of the reef Reef Egrets (*Egretta sacra*) silently stalked the shallows. By contrast, white surf thrashed the jagged seaward edge of the reef. At low water Haslewood and Lupton islands are joined by this reef, the largest in the Whitsundays. In a few hours the returning tide would bring with it a rich array of marine life, including reef fish, sharks and turtles.

The fringing reefs of the Whitsundays are the most varied and extensive in such close proximity to Australia's northern shores. They owe their existence to currents bringing coral larvae and other marine life from mid-shelf reefs into the limpid waters surrounding the islands and adjacent mainland. The most

diverse and elaborate reef formations are at the northern end of the group, such as those off Hook Island and Hayman Island.

These extraordinary structures are a major drawcard for visitors. They are also, arguably, the region's most precarious ecosystems. In addition to the destruction caused by natural events like cyclones, several of the reefs have been ravaged by sporadic infestations of the Crown of Thorns Starfish (*Acanthaster planci*). Bays like Manta Ray Bay on Hook Island and Cateran on Border Island are popular yacht anchorages and diving sites. Yet almost every time an anchor is dropped, more of the reef crumbles. It is estimated that around 60 per cent of the coral in these bays has been destroyed in recent years.

From White Bay I traversed the isthmus that tethers the bulk of Haslewood to its north-eastern promontory, Pallion Point. Within less than 100 metres I crossed from one arc of sand to another. And in Windy Bay the only things littering the beach were mangrove flowers and the tracks of Dotterels (*Charadrius* sp.). Skirting the water's edge I scrambled around rocky coves and headlands under the gaze of towering Hoop Pines. The bluffs were decked with garlands of Rock Orchids (*Dendrobium discolor*), their honey-coloured sprays vivid against the black basalt. Fish wagged through

Vivid honey-coloured sprays of Rock Orchids deck the bluffs bordering Windy Bay on Haslewood Island.

*A Green Turtle (*Chelonia mydas*), one of the many visitors to the coral shores of the Whitsunday islands.*

Branches of a Coastal She-oak frame the view from the white sand beach at Windy Bay on Haslewood Island.

the clear waters of the inlets, while overhead a pair of Ospreys (*Pandion haliaetus*) sailed on the warm breeze.

White Shoes, Whitehaven

The Whitsundays can be a place of baffling contradictions. Depending on where one stands—in both a physical and philosophical sense—the group can appear glorious and sublime or hopelessly sullied. It's a paradox I had been wrestling with ever since a visit to Hamilton Island some years earlier.

Formerly one of several islands leased for grazing, Hamilton was transformed during the 1980s into the region's largest resort. To some it is another monument to the excesses of the decade, an attempt to remodel a slice of the Whitsundays in the image of Waikiki. Certainly, it was a radical departure from the more homely resorts that had grown up on other islands in the decades since the area had first become popular for holidaymakers during the 1930s.

From our campsite on Haslewood Island we looked north-east, out across open water. A yacht would occasionally pass by on the horizon but the 'other side' of the Whitsundays remained out of sight. However, within five minutes of clambering aboard the Marine Parks boat *Wirrana* we rounded the point and headed for Whitehaven Beach on Whitsunday Island, back into tourism's mainstream.

Much of the 500 kilometres of island coastline in the Great Barrier Reef Marine Park is rocky and inaccessible. Only 25 kilometres of beach is considered suitable for recreation use. Whitehaven's beguiling six-kilometre-long strip of white sand is premium tourist real estate. As we approached the shore, luxury cruisers, yachts and tour boats all vied for space. Among them was the 'wave-piercing' catamaran bringing refugees from Hamilton Island for their dose of the 'real' Whitsundays. Inflatable dingies ferried day-trippers ashore to frolic in the pellucid shallows and along the sand, listening to the grains squeaking underfoot.

Whitehaven is also an area of extraordinary natural significance. The beach is 98 per cent silica, a relic of glacial times, when wind-blown

sand was driven against the eastern shores of Whitsunday Island. These formations, together with others on Haslewood Island, are the only dune ecosystems of their kind between Cooktown and Rockhampton. They are clad in an unusual mix of vegetation which includes the northernmost occurrence of Wedding Bush (*Ricinocarpos pinifolius*), together with communities of *Casuarina* and of *Acacia*, many of which are approaching senility due to dependence on fire for germination.

Whitehaven is one of the nesting and foraging areas for the vulnerable Beach Thick-knee (*Burhinus magnirostris*). The basalt headlands at the southern end of the beach are prime habitat for the Unadorned Rock-wallaby (*Petrogale inornata*).[5] Meanwhile, several species of turtles and the occasional Dugong (*Dugong dugon*) graze on the seagrass beds offshore.

Conserving such natural assets in the face of ever-increasing tourist pressures is a daily dilemma for rangers in the Whitsundays. Even the provision of basic facilities like toilets has become a major issue. With up to 500 people visiting Whitehaven at any one time, maintaining these 'facilities' is one of the most frequent assignments for park staff.

Even at present levels, the human impact on the islands is considerable. In addition to the 270 licensed 'bareboat' yachts and 350 commercial charter boats, there are many hundreds of privately owned vessels that use these waters every year. Moreover, it is estimated that by 2000, the annual rate of visitation will have increased two and half times. This, combined with a rapidly growing resident population on the adjacent mainland, will make the Whitsundays the busiest offshore region in Australia.

While rezoning and visitor management at sites like Whitehaven will help in the protection of the park, no amount of policing can completely control such a vast area where access is so unfettered. Ultimately, the best hope lies with a heightened public awareness of everyone's responsibility to the islands and their surrounding waters.

By far the greatest threat to the ecology, not just of the Whitsundays, but the entire coastal lagoon enclosed by the Great Barrier Reef, is the decline in water quality. Studies conducted in Repulse Bay to the south point to the crippling effect on coastal wetlands and reef communities caused by increased turbidity and nutrient levels, algal growth and sedimentation. Run-off from coastal rivers and land-based developments has seen a four-fold increase in sediment and nutrient levels in some offshore areas over the past 100 years.[6]

Leaving Whitehaven, we cruised north along the eastern shores of Whitsunday Island. While the talk on board was of coral core samples and dying reefs, it was reassuring to pass deserted coves and wild ridges. Then, rounding Pinnacle Point on Hook Island we motored into Manta Ray Bay where two yachts were at anchor. Here Mark Luyten, park ranger, toilet attendant and helicopter observer, assumed another of his guises by donning goggles and fins. As he slipped over the side he carried a stainless steel pin, part of a new fixed mooring for boats to tie up to. Paul would return the following day with a line and marker buoy. Given all the factors arrayed against the coral, discouraging boats from damaging these reefs is an urgent priority. As we left, another yacht headed into the bay. We rounded the corner and the boats disappeared behind a ridge. But we could hear the unmistakable sound of rattling chain and an anchor hitting the water.

1   J. Bates, *The Last Islands*, published by the author, 1990, p. 63.

2   G.L. Walsh, *Australia's Greatest Rock Art*, E.J. Brill/Robert Brown and Assoc., Bathurst, NSW, 1988, p. 126.

3   *Whitsunday National and Marine Parks Draft Management Plan*, Queensland Department of Environment and Great Barrier Reef Marine Park Authority, 1993, p. 16.

4   B. Barker, as quoted in *Whitsunday National and Marine*

Parks Draft Management Plan, Ibid, p. 31.

5   The other rock-wallaby species found in the area is on Gloucester Island, home to the only island-based population of the endangered Prosperpine Rock-wallaby (*Petrogale persephone*).

6   Dr John Brodie, Great Barrier Reef Marine Park Authority, as quoted in the *Sydney Morning Herald*, 18 February 1995.

WHITSUNDAY ISLANDS

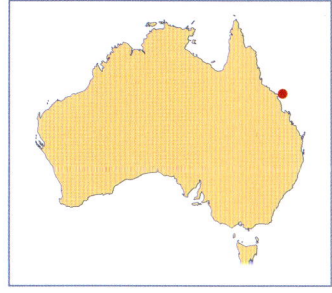

LOCATION

The Whitsunday Islands are dispersed off the Queensland coast between the towns of Bowen and Mackay. Hook Island 20°07' S, 148°55' E.

AREA

Whitsunday Island: 10,935 ha
Hook Island: 6,161 ha
Haslewood Island: 777 ha

CLIMATE

Subtropical/tropical

STATUS

With the exception of Hayman, Daydream, Titan, Hamilton and Dent islands, the Whitsundays form the Whitsunday Islands National Parks. This area is also encompassed within the Great Barrier Reef World Heritage Area. It is administered by the Qld Dept of Environment in conjunction with the Great Barrier Reef Marine Park Authority (GBRMPA).

ACCESS

Passenger boat services to Hayman, Hook, South Molle, Daydream, Long, Hamilton and Lindeman islands. Hamilton Island is serviced by commercial jet aircraft. Charter boats and yachts and water taxis are available from Shute Harbour and Airlie Beach. The waters surrounding the islands are subject to restrictions to protect the marine environment.

FACILITIES

Public campsites managed by the Queensland National Parks and Wildlife Service for commercial and private groups are found on the following islands: Gloucester, Saddlebank, Armit, Olden, Hook, Border, North Molle, Cid, Whitsunday, Haslewood, Tancred, Henning, Shaw, Thomas and South Repulse. Several resorts also offer camping facilities. Tourist resorts are located on Hayman, Hook, Daydream, South Molle, Long and Lindeman islands.

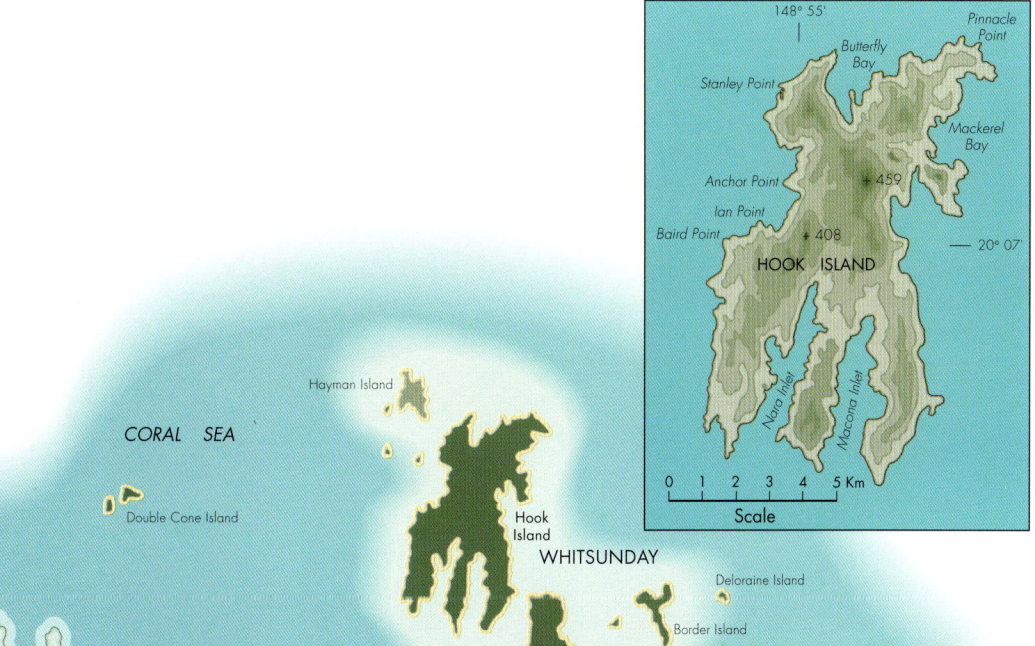

FRASER ISLAND

Storm Front

For 72 hours an intense low pressure system was stationed over the Queensland coast near Hervey Bay. A swirling mass of cloud brought thunder squalls and torrential downpours. Winds gusting in excess of 100 kilometres an hour whipped the sea into a tumult with five-metre waves pounding the shoreline. More than just a show of the elements, this storm was another episode in a long process of erosion and deposition. Churning through the surf were millions of tonnes of quartz grains, much of it bound for Fraser Island, the world's largest island made of sand.

Landing by barge during a lull in the storm, we found the island in turmoil. The ocean beach was abandoned. Clouds of salt spray and wind-driven sand streamed inland and piles of foam were strewn along the shore. In the forests we watched wind gusts shear leaves and branches from the canopy, while nearby the normally mirror-smooth lakes were furrowed by a frantic chop. Almost all the island's thousand or so visitors were huddled in camp grounds and resorts. For three riotous days nature triumphed over tourism.

In the pantheon of Australian islands none is more feted than Fraser. Each year thousands come to witness its immense dunes and coloured sand cliffs. Tour buses hurtle along the 123-kilometre surf beach to keep fleeting appointments with the forests and lakes, the freshwater streams and rocky headlands. Some arrive to pay homage to a wilderness saved. Others come simply to fish and swim and pitch their tents among the dunes.

Since the early 1970s the future of Fraser Island has been the focus of a torrid debate. The struggle to protect the island from sand mining involved competing interest groups and almost every strata of government and the judiciary. The process culminated in December 1992 with the inscription of Fraser Island as Australia's tenth entry on the World Heritage List. The island's natural assets were granted a reprieve, but the gales of protest had blown hard for 20 years. Now the island bears a heavy burden of popularity.

Grain Harvest

The splendours of Fraser Island have their origins 500 kilometres to the south in the mountains of northern New South Wales. Weathering of these granite and sandstone peaks produced vast quantities of sand sluiced into the sea by eastward-flowing rivers. A combination of ocean currents, wave action and prevailing south-easterly winds carried this sand north along the coast, until it eventually accumulated against prominent headlands and rock outcrops.

Over the past two million years these sand deposits have developed into massive dunes. During glacial periods, when sea levels dropped by nearly 100 metres, large expanses of sand were exposed and then mobilised by fierce winds to be deposited on the developing dunes. When the sea reached its present level 6,000 years ago, the upper dune formations of the Fraser Island and Cooloola sandmasses on the adjacent mainland were left high and dry.

On Fraser the dune structures extend to 100 metres below sea level and 240 metres above. Scientists have identified nine distinct phases of dune building on the island. Each of these overlapping accumulations of sand is associated with different successions of forest growth, soil development and dune erosion.[1]

Dunes continue to advance and retreat across the island. Each year an estimated 500,000 cubic metres of sand is transported along the Queensland coast. Some of this cargo is washed onto Fraser's shores and heaped onto coastal dunes by the south-easterly winds. This activity is not just a result of violent gales. Even 12 kilometre-an-hour breezes are sufficient to mobilise grains of sand. On almost any day of the year it's possible to watch the island take shape before one's eyes.

Sand finds myriad expression along Fraser Island's eastern shores, including dunes that, over the millennia, have become consolidated with coloured silts and clays. When exposed to the elements these ancient sand masses are abraded into fantastic forms.

One of the island's prime vantage points on the east coast is Indian Head, at the northern extremity of Seventy-five Mile Beach. In the wake of the storm we climbed to the summit of this headland. An exposed volcanic outcrop consisting of fine-grained rhyolite, it is one of only four small rock exposures on the island. In sharp sunlight the headland is a block of russet-brown moored between an azure ocean and a sea of sand. A pair of Brahminy Kites (*Haliastur indus*) wheeled above, and to the south a strip of surf and beach stretched unbroken to the horizon.

The accumulation of sand against this rocky bulwark was pivotal to the formation of Fraser. Eventually the same forces that built the dunes would usher in changes of another kind. As the nineteenth century unfolded, the onshore breezes brought not just drifting grains of quartz but the sails of north-bound ships.

Footprints

Indian Head was named by Captain James Cook as HMS *Endeavour* skirted this coast on 20 May 1770. He described it as a "black bluff head or point of land, on which a number of natives were assembled."[2] Cook's journal gives the first written account of this coast. Unaware that this shoreline was part of an island, Cook called it the Great Sandy Peninsula.

The first recorded European landfall was made by Matthew Flinders and crew members from HMS *Investigator* in July 1802. Flinders had first sighted Hervey Bay three years earlier on an exploratory voyage aboard the *Norfolk*. On his return visit he went ashore near Bool Creek on the north-western flank of Sandy Cape. With his Aboriginal companion, Bongaree, he made tentative approaches to "natives" congregated on the beach. Flinders observed that the Aborigines were well-built and appeared to thrive on the abundant seafood.

Based on his visits to Sandy Cape, Flinders conjectured that a passage connected the head of Hervey Bay with the open sea. This was eventually confirmed in 1822 when Captain William Edwardson navigated the Great Sandy Strait in the cutter *Snapper*. On the basis of this foray he noted that "the natives were numerous and hostile."[3] In subsequent decades such suspicions were inflamed by repeated clashes between European settlers and dispossessed Aborigines, as well as stories like those surrounding an Englishwoman cast onto the shores of the island that later took her name.

The brig *Stirling Castle*, en route to London via Singapore, was wrecked on the Swain Reefs on 22 May 1836. Hoping to reach Moreton Bay, the stronger crew members manned the longboat. The remaining survivors, including the ailing Captain James Fraser and his pregnant wife, Eliza, were towed in the ship's pinnace. It was a horrendous journey: both boats leaked and fresh water was in short supply. During the voyage Eliza gave birth, but the child did not survive. After two weeks of slow progress the frustrated longboat crew cut the pinnace loose. In time, the Frasers and five others landed near Waddy Point on 26 June 1836.

According to Eliza's narratives, her husband was subsequently speared to death by Aborigines and she was held "captive" and "treated with the greatest cruelty." After seven weeks in the company of the Badtjala people she was eventually rescued by a search party sent from Moreton Bay under the leadership of Lieutenant Charles Otter. The truth of her ordeal soon became blurred. Lurid accounts of "dreadful slavery, cruel toil and excruciating tortures"[4] were published after her return to England—tales that fed a growing prejudice towards Aboriginal people.[5]

Yet there was another side to this story. Among those instrumental in Otter's rescue of Eliza were two escaped convicts, John Graham and David Bracewell. Fleeing the terrors of Moreton Bay, they were among many 'white blackfellows' or bunders who went bush, often for years at a time, taking their chances among the Aborigines. It was Graham and Bracewell's knowledge of the country and people on and around Fraser Island that made Eliza Fraser's rescue possible.

Dispossession

Six years later Andrew Petrie, an official from the Moreton Bay penal settlement, visited the

Above: Strangler Figs, whose entwining roots clench and ultimately engulf their host trees, are a common sight in the island's tall forests.

Opposite: Along clear-flowing creeks and moist gullies, a bewitching array of rainforest plants—including tall Piccabeen Palms and the giant King Fern—vie for their ration of precious sunlight.

area, ostensibly to look for Captain Fraser's grave. He found no trace of the hapless captain but he did bring back glowing reports of the island's spectacular forests. This discovery, even more than the melodrama of Mrs Fraser, was to have a devastating impact on what was one of the richest and most densely populated regions in Aboriginal Australia.[6]

At the time of European contact three main Aboriginal groups occupied Fraser Island. The Ngulungbara lived in the north around Sandy Cape, the Badtjala resided in the centre of the island, while the Dulingbara held the country in the south. Collectively they were known as the Badtjala people and their social relationships and tribal laws were as complex and vigorous as the environment they inhabited. During the winter months, when the sea was most productive, 2,000–3,000 people occupied the island. As well as gathering oysters and molluscs, they hunted fish, turtles and Dugong in the sheltered waters of Hervey Bay.

From the forest trees the Badtjala people cut great sheets of bark with which they built canoes and roofed their *gunyah* shelters. Thin bark strips were woven into fishing nets and pandanus leaves fashioned into carry baskets. Their diet included yams, fern roots, grass tree leaves, cycad fruit and assorted smaller fruits. Native bees were a source of honey, and the wax was used for myriad purposes. Over thousands of years the Badtjala had developed an intimate relationship with the island that was the embodiment of their creation spirit, K'gari.

By the mid-nineteenth century the violence between European settlers and Aboriginal groups around Maryborough had spread to the island. Fraser was gazetted as an Aboriginal reserve in 1860 but this was revoked two years later with renewed interest in the forest timber. The first Kauri Pine (*Agathis robusta*) was felled by "Yankee Jack" Piggot in 1863 and punted across to the mainland sawmill set up by entrepreneur William Pettigrew. The following year a Badtjala spear felled Piggot himself—most likely a reprisal for his mistreatment of the tribe.

In the space of four decades the combined effects of massacres, dispossession and disease reduced the island's population from 2,000 to a mere dozen or so. In 1905 Archibald Meston,

the Queensland Protector of Aborigines wrote: "A few old survivors are scattered over the mainland and some of their descendants haunt the vicinity of Maryborough, mournful, opium-poisoned parodies of their healthy, free and warlike ancestors."[7] Meston and others tried to "ease the plight" of the Badtjala people by establishing missionary settlements. By 1906, however, the last Aboriginal reserve on the island had been revoked. White man's law had severed the 5,500-year-old relationship between K'gari and the Badtjala people.

Silicates To Satinays

By the turn of the century the forests of the island no longer echoed to corroboree songs but to the sounds of axe blows and cross-cut saws. At first the timber-getters went after White Beech (*Gmelina leichhardtii*) and Kauri Pine. Later they also took other species, including Tallowwood (*Eucalyptus microcorys*), Blackbutt (*Eucalyptus pilularis*) and Brushbox (*Lophostemon confertus*). The grandest of their targets was Satinay (*Syncapia hillii*)—1,000-year-old specimens, some which stood 60 metres high with trunks three metres in diameter.[8] By 1948 nearly 70 percent of the island's original timber reserves had gone.

The existence of such forests on an island composed almost exclusively of sand is but one of Fraser's ecological wonders. An extraordinary mosaic of plant communities, including at least 625 native species, have colonised the sandmass. On a journey across the centre of the island from east to west it is possible to traverse a sample of these communities.

The island's east coast is a harsh place. Lashed by salt spray and ocean winds, only the hardiest plants survive on the exposed, sun-baked foredunes. An array of salt-resistant vines and succulent creepers colonise this frontline. Plants like Goatsfoot Convolvulus (*Ipomea pescaprae*), Angular Pigface (*Carpobrotus glaucescens*) and Beach Spinifex (*Spinifex sericeus*) help bind the sand and concentrate nutrients in the topsoil.

Close behind these tenacious ground-huggers stand Beach Oaks (*Casuarina*

One of the island's prime vantage points on the east coast is Indian Head, at the northern extremity of Seventy-five Mile Beach. In the wake of the storm we climbed to the summit of this headland. An exposed volcanic outcrop consisting of fine-grained rhyolite, it is one of only four small rock exposures on the island. In sharp sunlight the headland is a block of russet-brown moored between an azure ocean and a sea of sand. A pair of Brahminy Kites (*Haliastur indus*) wheeled above, and to the south a strip of surf and beach stretched unbroken to the horizon.

The accumulation of sand against this rocky bulwark was pivotal to the formation of Fraser. Eventually the same forces that built the dunes would usher in changes of another kind. As the nineteenth century unfolded, the onshore breezes brought not just drifting grains of quartz but the sails of north-bound ships.

Footprints

Indian Head was named by Captain James Cook as HMS *Endeavour* skirted this coast on 20 May 1770. He described it as a "black bluff head or point of land, on which a number of natives were assembled."[2] Cook's journal gives the first written account of this coast. Unaware that this shoreline was part of an island, Cook called it the Great Sandy Peninsula.

The first recorded European landfall was made by Matthew Flinders and crew members from HMS *Investigator* in July 1802. Flinders had first sighted Hervey Bay three years earlier on an exploratory voyage aboard the *Norfolk*. On his return visit he went ashore near Bool Creek on the north-western flank of Sandy Cape. With his Aboriginal companion, Bongaree, he made tentative approaches to "natives" congregated on the beach. Flinders observed that the Aborigines were well-built and appeared to thrive on the abundant seafood.

Based on his visits to Sandy Cape, Flinders conjectured that a passage connected the head of Hervey Bay with the open sea. This was eventually confirmed in 1822 when Captain William Edwardson navigated the Great Sandy Strait in the cutter *Snapper*. On the basis of this foray he noted that "the natives were numerous

and hostile."[3] In subsequent decades such suspicions were inflamed by repeated clashes between European settlers and dispossessed Aborigines, as well as stories like those surrounding an Englishwoman cast onto the shores of the island that later took her name.

The brig *Stirling Castle*, en route to London via Singapore, was wrecked on the Swain Reefs on 22 May 1836. Hoping to reach Moreton Bay, the stronger crew members manned the longboat. The remaining survivors, including the ailing Captain James Fraser and his pregnant wife, Eliza, were towed in the ship's pinnace. It was a horrendous journey: both boats leaked and fresh water was in short supply. During the voyage Eliza gave birth, but the child did not survive. After two weeks of slow progress the frustrated longboat crew cut the pinnace loose. In time, the Frasers and five others landed near Waddy Point on 26 June 1836.

According to Eliza's narratives, her husband was subsequently speared to death by Aborigines and she was held "captive" and "treated with the greatest cruelty." After seven weeks in the company of the Badtjala people she was eventually rescued by a search party sent from Moreton Bay under the leadership of Lieutenant Charles Otter. The truth of her ordeal soon became blurred. Lurid accounts of "dreadful slavery, cruel toil and excruciating tortures"[4] were published after her return to England—tales that fed a growing prejudice towards Aboriginal people.[5]

Yet there was another side to this story. Among those instrumental in Otter's rescue of Eliza were two escaped convicts, John Graham and David Bracewell. Fleeing the terrors of Moreton Bay, they were among many 'white blackfellows' or bunders who went bush, often for years at a time, taking their chances among the Aborigines. It was Graham and Bracewell's knowledge of the country and people on and around Fraser Island that made Eliza Fraser's rescue possible.

Dispossession

Six years later Andrew Petrie, an official from the Moreton Bay penal settlement, visited the

Besieged by an advancing dune moving inland by as much as a metre a year, Lake Wabby is graphic testimony to the mercurial moods of Fraser Island's sandscapes. Quentin Chester

Lilies (Nymphaea sp) are a striking sight in freshwater soaks along the coastal hinterland.

equisetifolia). Other conspicuous dune-dwellers include the Coast Banksia (*Banksia integrifolia*) and Pandanus (*Pandanus tectorius*).

The build-up of leaf-litter and increased fertility in the lee of the foredunes encourages a more diverse association of shrubs, often interspersing groves of Coastal Cypress (*Callitris columellaris*) and Paperbarks (*Melaleuca quinquenervia*).

Further inland the vegetation cover gradually develops, commonly forming a low sclerophyll forest, including Scribbly Gums (*Eucalyptus signata*) and Moreton Bay Ash (*Eucalyptus tessellaris*). This open woodland gives shelter to smaller wattles, She-oaks and cycads. Large, flat expanses are carpeted by shrubby heathland plants, notably the Wallum Banksia (*Banksia aemula*) and *Banksia oblongifolia*. The nectar of the banksia flowers is a lure to the Queensland Blossom Bat (*Synconycteris australis*), the smallest of the herbivorous fruit bats. These sprawling heathlands also form the northernmost habitat in eastern Australia of the rare Ground Parrot

(*Pezoporus wallicus*).

On high ridges in the heart of the island tall forests of Blackbutt encircle sheltered valleys formed by the ancient dunescape. Sequestered in these fertile, well-watered havens are the rainforest giants. Towering specimens of Satinay, Brushbox, Kauri Pine and Hoop Pine (*Araucaria cunninghamii*) merge with the stately Piccabeen or Bangalow Palms (*Archontophoenix cunninghamiana*) and an understorey of Carrol (*Backhousia myrtifolia*) to form a lush canopy.

Within these cloistered realms, vines, orchids and epiphytic ferns flourish. The canopy is a haunt of the Grey Headed Flying Fox (*Pteropus poliocephalus*), a fruit and blossom feeder. Meanwhile the forest floor is patrolled by numerous reptiles and small mammals, including the Southern Bush Rat (*Rattus fuscipes*) and the Fawn-footed Melomys (*Melomys cervinipes*).

Rainforests are the zenith of Fraser Island's plant succession. And yet the tenure of the forests is not indefinite. Prolonged weathering gradually leaches the nutrients out of the reach of root systems, submerging the nutrients as much as 25 metres below the surface. Stunted woody heaths and wetlands on the western side of the island are evidence of impoverished soils on Fraser's oldest dunes. The ridges are clad in Swamp Banksias (*Banksia robur*) and mallee eucalypts. Stands of Coastal Cypress and Paperbark occupy low-lying areas and adjoin large areas of swamp and grass land.

The transitions across Fraser's botanical time scale can happen in the space of a few metres. One moment one is in open woodland of Banksias and Grass Trees, the next in an enveloping forest of 20-metre-high Blackbutts. It is as though the island's flora has been stage-managed to showcase as many congregations of plants as possible, dealing with everything from salinity to senility, all in a 20-kilometre stretch.

Lakes District

In the depths of the rainforests this sense of juxtaposition is all-embracing. One is confronted by the colossal girth of canopy trees alongside the fine brocade of pendulous ferns, or delicate orchids against the Strangler Fig's (*Ficus watkinsiana*) writhing limbs. Where an aging giant has fallen and the canopy is breached, dazzling light blazes to the ground.

Young paperbarks colonise the shallow margins of Lake Boomanjin, the world's largest perched lake.
Rob Jung

Nectar from the Banksia
serrata, *a signature species of
the island's woodlands, is
prized by many creatures,
including the White-cheeked
Honeyeater (*Phylidonyris
novaehollandiae*).
Rob Jung

Then a few metres away arching fronds layer shadow upon shadow.[9] The convulsions of the weather during our visit only served to heighten this air of unreality. During squalls, rain clattered down onto the outstretched foliage of Elkhorns (*Platycerium bifurcatum*) and Bird's-nest Ferns (*Asplenium australasicum*). In moist gullies King Ferns (*Angiopteris evecta*) gleamed with a lacquering of new moisture.

If wind is the island's progenitive force, then water is the great provider, with the power to foster some of world's most resplendent forests and then ultimately starve them out of existence. Average annual rainfall ranges from 1,250 millimetres at Sandy Cape to more than 1,750 millimetres on the higher dunes. The island is remarkably absorbent. Rainfall soaks directly into the sand, where it becomes trapped in an immense aquifer containing many billions of litres of water, forming Australia's largest single reservoir of potable water. The water-table forms a dome-like curve within the sandmass, rising more than 150 metres above sea level and nearly 40 metres below. On average, water remains trapped in the aquifer for 100 years before spilling into the sea via underground streams and coastal creeks, mainly on the western shores. The largest outflow comes from Bogimbah Creek, which daily disgorges 166 million litres of pure, filtered water into Great Sandy Strait.

Dotted across the island are more than 40 freshwater lakes. Where depressions in the dunes dip below the watertable, window lakes are formed. These limpid ponds are literally a window to the subterranean aquifer. Other lakes are elevated and owe their existence to organic material compressed to form a hard layer of impermeable humate lining shallow dune depressions. Fraser has the world's greatest collection of these perched lakes, including the largest, Lake Boomanjin (200 hectares), and the highest, the Boomerang Lakes (130 metres above sea level). The Wabby Lakes are formed by an encroaching dune building a dam wall against a small stream to create the island's deepest lake (11.4 metres).

Fraser's lakes are tranquil, reflective places. Two days after the storm we were camped on the shores of Lake Boomanjin. This ancient crater-like depression is ringed by a ridge of eucalypt and banksia scrub. Prevailing south-easterly winds have created a broad beach backed by dunes on the lake's north-western shores. This arc of brilliant white sand is braided with thin streams in hues of amber and deep orange—organic tannins that give the lake the colour of stewed tea.[10]

Boomanjin's eastern edge is fringed by sedges and gnarled Paperbarks whose roots twine through the stained banks and crusts of 'coffee-rock', ancient beds of sand compacted with humus. In the adjoining scrub, wind-sheared branches and a confetti of freshly fallen leaves lies waiting for the agents of decay that will in time transform this litter into humus. While evidence of the recent gales was strewn around, out on the lake all was still, the only sound a chorus of frogs. It was as though the storm had never happened.

Feathers, Furs, Fins And Ferals

Like most lakes on the island, the sand-filtered waters of Boomanjin are acidic and low in nutrients. These harsh conditions support only a limited fauna, including amphibians adapted to the acid waters. The most engaging lake occupants are freshwater turtles. Of the island's three species the Kreft's River Turtle (*Emydura kreftii*) is the most common and, like many of the island's freshwater creatures, its diet includes an array of insect life. [11]

While the relative paucity of life in the lakes attracts few waterbirds, other habitats are well endowed. Approximately 300 species are found in the region. An array of whistlers, flycatchers and honeyeaters inhabit the grassy swathes and open woodland. The island is also a refuge for rare and endangered species such as the Brush Bronze Pigeon (*Phaps elegans*), Black-breasted Button-quail (*Turnix melanogaster*) and the Marbled Frogmouth (*Podargus ocellatus*).

The Great Sandy Strait that separates Fraser from the mainland is one of the largest and least disturbed estuarine habitats in southern Queensland. The island's coastal margins support one species of mangrove. Swamp She-

*The Fraser Island region is acknowledged as one of the top three summer stopovers in Australia for trans-equatorial migratory wading birds such as this Bar-tailed Godwit (*Limosa lapponica*).*
Rob Jung

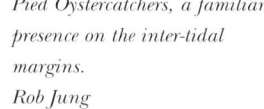

Pied Oystercatchers, a familiar presence on the inter-tidal margins.
Rob Jung

oaks (*Casuarina glauca*) have also extended their range into the inter-tidal zone. This vast tract of saltmarshes, mangrove flats and sandy spits is one of Australia's three most significant summer stopovers for migratory wading birds.[12]

One hundred thousand hectares of sea grass carpets the seabed of the strait and Hervey Bay, forming the largest beds in eastern Australia. This expanse is a major grazing area for Dugong (*Dugong dugon*).[13] Six species of turtle are found in these waters, and Hervey Bay is the most significant breeding area in the south-west Pacific for the endangered Loggerhead Turtle (*Caretta caretta*).[14] These sheltered waters are also an important stopover for the endangered Humpback Whale (*Megaptera novaeangliae*). From August to October several hundred whales, including mothers with calves, rest in the bay during their migration from their tropical breeding grounds.

For all the profusion of marine life in the surrounding waters, Fraser's 184,000 hectares support surprisingly few large native animals. Most of the 21 indigenous mammals are small creatures such as bandicoots, marsupial mice and native rodents. There are no emus, kangaroos, koalas or scrub turkeys, and only minor populations of wallabies and possums—

possibly due to a lack of suitable food sources, though Badtjala hunters most likely accounted for the demise of some species.

More recent arrivals are also now making their impact. Dingoes (*Canis familiaris*) were originally brought to the island by Aborigines. The Fraser population, the purest strain of eastern Australia, is now itself at risk from feral dogs and the deadly *Parvo* virus. Feral cats are major predators of small mammals. Meanwhile Brumbies, descendants of horses introduced by the timber-getters in the 1870s, trample vegetation and erode fragile foredunes.

Sand Highways

The most conspicuous fauna on the island are introduced humans ("*Homo touristus fraseri*"). Transitory groups visit the island in long-awaited migrations, mainly from strongholds in the south-east of the continent. They arrive by barge, passenger ferry and light aircraft to roam the island and browse on its scenic wonders. Since the late 1960s, when the annual influx was around 5,000, this pilgrimage has grown to 350,000-strong.

Camped on Seventy-five Mile Beach we had ample opportunity to observe the behaviour of

The Humpback is one of the most exuberant of whale species. Each spring several hundred Humpbacks visit the sheltered waters of Hervey Bay during their long journey south to Antarctic feeding grounds.

our own kind. With the ebbing tide the beach became littered with tiny sand pellets created by many thousand Sand Bubbler Crabs (*Scopimera inflata*) sifting the moist grains for food particles. Nearby, pairs of Pied Oyster Catchers (*Haematopus ostralegus*) speared the fleshy meat from shells. But these sights were overtaken by an erratic procession of vehicles that repeatedly scattered Crested Terns (*Sterna bergii*) trying to congregate on the sand. For much of the day the spectacle of the Pacific

Ocean thumping onto the shore was lost in a blur of Pajeros and Land Cruisers. By the time the tide began to return, the sweep of sand was scored by tyre tracks—including our own.

The sheer volume of visitors to Fraser poses a substantial threat to the island and its frangible sandscapes. Dune erosion, litter, degraded campsites, lake pollution and track damage— all are plain to see. Effective management of the island is further complicated by legacies from the recent past, including resort

Scribbly-gums, banksias (Banksia robur pictured) and Grass Trees (Xanthorrhoea spp) are characteristic of Fraser Island's most widespread woodland plant communities. Rob Jung

developments, a network of forestry roads, and makeshift settlements. There is even a private housing estate. But perhaps the most intractable legacies are the competing expectations among the island's visitors.

By late afternoon the traffic on the beach resembled suburban rush-hour as convoys rumbled past, racing to fishing spots and campsites in the north. Fraser may have been spared the physical ravages of sand mining but according to some onlookers, the soul of the island is being quarried by tourism. Feeling overwhelmed by the view along the beach, I headed inland and withdrew into the sanctity of an immense valley—one of Fraser's prodigious sandblows.

Making Tracks

The eastern margins of the island are repeatedly invaded by these glaciers of sand. They originate where breaches in the vegetation allow onshore winds to excavate frontal dunes, forming deep U-shaped valleys, with an advancing wall of sand smothering any vegetation in its path. These parabolic dunes loom as examples of the havoc caused by erosion—both in the natural course of events and through wanton disturbance by man.

The sandblows are wonderful, forbidding places. Just metres from our camp I entered a huge scoop of bare sand flanked by 50-metre-high walls. The sounds of the surf and traffic were completely muffled. From the rippled floor of the blow rose the skeletons of trees, interred centuries before and now exhumed by the winds. Nearby stood mounds of bleached shells, left over from Badtjala feasts in the distant past. I climbed high onto the leading edge of the blow, where the dune headwall was engulfing a grove of cypress and banksia, their limbs being dragged into a suffocating drift of golden sand.

In this austere domain there was complete seclusion. Lying on the crest, its surface stippled from the recent rains, I sank my fingers into the dune, past the dry, outer crust to the cool, moist grains below. In my solitude the clamour of the beachfront seemed a world away. And yet the vanquished forest loomed as an omen. Ambling back to camp, I looked back at the great sweep of sand disfigured by my plunging footsteps and was grateful for the gritty wind that raced past at ankle height and began covering my tracks.

1   *Great Sandy Region—Draft Management Plan*, Queensland Department of Environment and Heritage, 1993, pp. 37–9.

2   J.C. Beaglehole, *The Journals of Captain James Cook on His Voyages of Discovery*, vol. 1, Hakluyt Society, Cambrige, 1955.

3   As quoted in F. Williams, *Written In Sand*, Jacaranda Press, Sydney, 1982, p. 27.

4   From the title page of *The Shipwreck of Mrs Fraser*, Dean and Murray, London, 1838.

5   Over the decades aspects of Eliza Fraser's story have become enshrined in popular mythology, most notably through an eponymous feature film, a series of paintings by Sidney Nolan and Patrick White's novel, *A Fringe of Leaves*.

6   Dr N. Tindale, as quoted in J. Sinclair, *Fraser Island and Cooloola*, Weldon Publishing, Sydney, 1990.

7   A. Metson, *Report on Fraser Island*, Queensland Legislative Assembly, 1905, quoted in Williams *op. cit.*

8   The resistance of the Satinay to attack from marine borers saw it put to use in sites such as London's Tilbury Docks and the Suez Canal. Logging continued on a smaller scale until 1993.

9   P. White, *A Fringe of Leaves*. ". . .the trees were so densely massed, the columns so mass upholstered or lichen encrusted, the vines suspended from them so intricately rigged, the light barely slithered down . . ."

10   Such 'black' water is common in the perched lakes, though Lake Mackenzie and Lake Birrabeen are examples of perched lakes with where filtration of organic material by adjacent dunes has created 'white water'.

11   Boomanjin is home to the world's most primitive Chironomid (*Diptera*), a kind of tiny midge. This recently identified genus is a member of a sub-family not previously found in Australia.

12   The strait is also considered Australia's single most important habitat for the Eastern Curlew (*Numenius madagascariensis*).

13   In 1988 their population was estimated at 1,800 but a large-scale depletion of the seagrass in 1992 saw that figure drop to around 200.

14   *Great Sandy Region—Draft Management Plan*, Queensland Department of Environment and Heritage, 1993, p. 55.

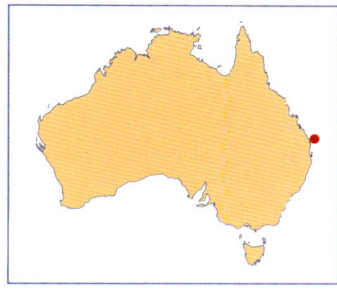

LOCATION

25°15' S, 153°15' E. Adjacent to Hervey Bay, 190 km north of Brisbane.

AREA

184,000 hectares

CLIMATE

Subtropical

STATUS

Fraser Island World Heritage Area is managed by the Queensland National Parks and Wildlife Service. The island includes freehold and leasehold land as well as township reserves.

ACCESS

Passenger launches and vehicular barges operate from Inskip Point, River Heads and Urangan. Charter Flights are also available. Four-wheel-drive vehicles are essential for travel around the island.

FACILITIES

Townships at Eurong and Happy Valley. Campsites at Central Station, Lake Boomanjin, Lake Mackenzie, Lake Allom, Dundubara, Waddy Point and Wathumba. Resorts are located at Dilli Village, Eurong, Happy Valley, Cathedral Beach and Kingfisher Bay.

153° 00'

Sandy Cape

Sandy Cape Lighthouse

Rooney Point

MARLOO BAY

PLATYPUS BAY

Waddy Point

25° 00'

Wathumba

Indian Head

HERVEY BAY

0   5   10   15   20 Km

Scale

Lake Bowarrady

Allom Lake

The Cathedrals

Boomerang Lakes

Moon Point

Hervey Bay

Eli Creek

FRASER ISLAND

CORAL SEA

Lake McKenzie

Lake Wabby

Central Station

Maryborough

Lake Birrabeen

Lake Benaroon

25° 30'

Lake Boomanjin

Seventy Five Mile Beach

Great Sandy Strait

Hook Point

Inskip Point

Rainbow Beach

Animal Architecture

Night had fallen over the lagoon. The MV *Tamaru* came to anchor within a cordon of masthead lights and dancing reflections more than 50 kilometres from the closest point on the Queensland coast. Although surrounded by the open Coral Sea, any swell was dampened by encircling arms of coral as *Tamaru* rode peacefully over sand just seven metres below her keel. As the moon emerged from the sea, the dark profile of Lady Musgrave Island stretched out on a far corner of her own glassy lagoon.

East of Gladstone, Lady Musgrave Island and its attendant reef are the second-most southerly on the Great Barrier Reef, small outliers of the largest structure ever fashioned by any life form or group of organisms. An intricate chain of about 3,400 individual coral reefs extends from near Bundaberg, north almost as far as Papua New Guinea. Through most of its length, the reef runs parallel with the Queensland coast, fastened to the outer submarine edge of the continental shelf. Tiny Lady Elliot Island, at the reef's southern extremity is 2,300 kilometres from its northern limits and only 30 kilometres from Sandy Cape on Fraser Island.[1] This is animal architecture on a stupendous scale. Coral polyps, incalculable billions of tiny marine creatures, have built most of this great limestone mansion.

Among the strands and swirls of reef, coral islands known as cays linger just above the tide, cast across the seascape like miniature cameos in settings of baroque extravagance. Many are no more than banks of sand and coral rubble thrown up on the reef platform by storms and

Opposite: Wreck Island lies within the Great Barrier Reef Marine Park, the world's largest. Management of this vast area attempts the Solomon-like task of conserving a great natural phenomenon, while catering for commercial and tourist interests. Wreck and its surrounding waters see none of these. They are preserved exclusively for their natural values, set aside as a pristine and priceless benchmark. AlasdairMcGregor/Qld Dept of Environment

Below: Hoskyn Islands (pictured from the north) and nearby Fairfax Islands in the Bunker Group are examples of rare double cays. Bob Mossel

Behind the sandly littoral fringe, Octopus Bush hedges the vegetated centre of Hoskyn Islands' western cay. On the eastern, weather side of the lagoon a second cay has formed composed largely of shingle. Ocean swells and cyclones provide the energy to sweep up coral fragments to form shingle cays.

ocean swells. Others have withstood the restlessness of the elements for long enough to capture freshwater and colonising plants.

The cays of Capricornia lie on average about 80 kilometres from the mainland. Nine in the Capricorn Group, three in the Bunker Group (including Lady Musgrave), together with isolated Lady Elliot Island, are separated from the rest of the Great Barrier Reef by more than 100 kilometres of open sea across the

Capricorn Channel. With no protective outer "barrier" each island is set on its own platform reef, one of a total of 21 roughly palette-shaped abstractions of sumptuous colour.

The fanciful light-show playing on the lagoon signalled the presence of at least 20 yachts and cruisers anchored within its confines. In addition to the boats, the island was a temporary home to about 50 campers. September school holidays were in full swing

Growth and Decay

According to the locals, it was *that* time of year. The wind hunted restlessly around the compass, not settling to any particular pattern. One thing was certain, though: whether from the south-east or north-west it generally blew hard — up to 20 or 25 knots on some afternoons. But the prevailing conditions were of no concern to Crested and Black-napped Terns (*Sterna bergii* and *S. sumatrana*) as they loafed together in a chattering bunch at the end of a sand-spit on tiny Erskine Island. A strong north-westerly pushed a choppy shore break across the narrowest section of the reef-flat. The waves momentarily stood up, opalescent in the afternoon sun, then thwacked into the steep face of a rapidly eroding sandbank. If the wind persisted and strengthened, it seemed as though the island's narrow ring of sand would be washed away.

Looking past the reef crest to nearby Masthead Island, floating like a raft on an argentine sea, I was struck by the creative tension of the place. A perpetual struggle between the ephemeral and the immutable is being waged around these islands and their parent reefs.

The Capricornia cays are but geological infants. They are perhaps just 6,000 years old, forming only after the sea level rose and stabilised at the end of the last ice age. Twelve thousand years before, the shallow seabed that is now the continental shelf was a dry coastal plain clothed in eucalypt woodland. Aborigines hunted game, roaming its low limestone hills in an area where turtles swim today. These hills were the eroded remnants of more ancient reef systems, consigned to the geological record by fluctuations in sea level during the previous two million years.

As the sea again began to rise about 10,000 years ago, the coastal lowlands were slowly inundated. The highest points remained above the tide, creating the 600 or so continental or "high" islands scattered along the coast. In the surrounding warm shallow waters juvenile coral reefs flourished, growing ever higher as the sea level swelled. Today's growing coral is only a thin living "skin" stretched over the body of the mature reef. Hundreds of metres of limestone

and Lady Musgrave was as popular as ever. Our arrival was deliberately timed. I was accompanying a team of Marine Parks officers on a patrol of the entire length of the Capricorn Bunker Group. Their tasks were as varied as the aspirations of those that visit these waters. Education, interpretation, campground and wildlife monitoring, law enforcement and the collection of bizarre flotsam from deserted beaches — it was all in a day's work.

In shallow, clear, nutrient-poor tropical waters, minute algae form the solar cells of the coral reef. They capture energy through photosynthesis and exchange food and wastes for shelter within the bodies of their hosts, the coral polyps. Their presence enhances the polyps' vigour and the rate at which calcium carbonate is extracted from sea water. Simple corals are thus made reef-builders.
Greg Carter

top the reef crest and push across their tabletop flats. Rather than being randomly scattered, debris and sediments are swept into a tidy pile, usually towards the leeward edge of the reef. This heap of the reef's own detritus — the incipient cay — is quite unstable and may easily be washed away by storms or cyclones. And so the process starts again.

"... Bunker's Isles ..."

In 1770, James Cook became the first European navigator definitely known to have sailed within the confines of the Great Barrier Reef.[2] Cook had already voyaged 1,000 kilometres past the most southerly beginnings of the perilous maze before HMS *Endeavour* was forced to negotiate a passage through coral. His route had kept to inshore waters, where he neither saw nor suspected what flanked the *Endeavour* to the east and north.

Like Cook, Matthew Flinders hugged the coast in 1802 as he sailed north on the first leg of HMS *Investigator*'s circumnavigation of the continent. Flinders also missed the cays at the southern end of what he came to call "the Barrier Reefs". The string of reefs and islands that straddle the Tropic of Capricorn and have collectively come to be known as "Capricornia", were first sighted in 1803 by Captain Eber Bunker aboard the whaling vessel *Albion*.

Phillip Parker King encountered and named Lady Elliot Island in 1819 on the first of his forays into reef waters and made further reference to "Bunker's Isles" during his coastal surveys of 1820 and 1821. He observed that "they abound with turtle and bêche-de-mer, the latter of which, if not both, will at some future time become of considerable importance to the coasting trade of New South Wales".[3]

The first scientific surveys of the group occurred when HMS *Fly*, accompanied by HMS *Bramble*, visited Capricornia in January 1843 under the command of Francis Blackwood. The two ships were at the start of a four-year survey of the Great Barrier Reef, Torres Strait and New Guinea. The account of Professor J. Beete Jukes, geologist and naturalist on the expedition, gives a vivid impression of a pristine reef environment and coral cays before human

lie beneath, deposited as the skeletal remains of innumerable generations of hard coral polyps among more than 350 species.

The cays are comprised of dead fragments of this skin in the form of coral rubble and sand, broken and brushed off the reef by wind and wave. Great quantities of sand are also made by the calcerous remains of encrusting algae and the shells of untold numbers of marine animals. In a series of complex interactions between shape, tide and orientation, some reefs manage to refract and bend waves as they disturbance.

Jukes named Heron Island for the numerous Eastern Reef Egrets or Reef Herons (*Egretta sacra*) that nested there, and observed seabirds and turtles in great abundance throughout the group, writing that "one night Lieutenant Shadwell, on one of the islands observing star altitudes, was actually obliged to place sentries around him to prevent the turtles from running over his artificial horizon as it lay on the ground".[4]

Jukes and William Saville-Kent, who spent time on Lady Elliot, Heron and North West Islands in the 1890s, had to content themselves with observations of the coral environment at low tide on the reef flat. The thrill of underwater exploration with the aid of scuba was then many years away. Thus Jukes's observation of the drying reef at One Tree Island was one of disappointment: "There was no signs of living coral, except a few stunted specimens in some of the deeper holes of the reef ... It looked simply like a half drowned mass of dirty brown sandstone..."[5]

From Turtle Soup to Turtle Tags

By the time Saville-Kent visited the Capricorn Bunker Group, exploitation of its resources had been underway for more than 75 years. Commercial harvesting of the Great Barrier Reef began in 1804 after an enterprising Captain James Aickin investigated Wreck Reef, 350 kilometres east of Shoalwater Bay, far out in the Coral Sea. Aickin salvaged brass and iron from the wreck of the *Porpoise*, on which

A Loggerhead Turtle (Caretta caretta) digs her nest by night in soft Heron Island sand. In addition to several Capricornia cays, the critically endangered Loggerhead nests on the mainland near Bundaberg. Wreck Island is their most important nesting site in the South Pacific. As many as 1,000 females may use the island in a single summer.

Matthew Flinders had been attempting to sail to England.[6] But Aickin saw a more lucrative prize than scrap metal in the echinoderms inhabiting the shallows of the reef.

Seeking a more accessible source of what was commonly known as bêche-de-mer, or trepang,[7] Aickin set up operations on Lady Elliot Island. Easily collected on the reef-flat, a cargo of boiled, dried and salted bêche-de-mer was taken to Sydney after only three months. Prized by the Chinese, it was then shipped to Canton. Bêche-de-mer was one of the colony's earliest exports. By the 1880s bêche-de-mer fishermen were working through the length of the Great Barrier Reef and in 1882 bêche-de-mer worth over £25,000 was exported from Queensland.

Guano miners were next. It is unclear when mining actually began but two ships, the *Bolton Abbey* and *Countess of Minto*, were sent from Sydney to Lady Elliot to collect guano in 1851. The venture was unsuccessful, with one ship wrecked on the island and the other swept out to sea in a storm. The Anglo-Australian Guano Company was established in Hobart in the early 1860s and began operations on Lady Elliot Island in 1863 under a lease agreement with the Queensland Government.[8] A workforce of Indian, Chinese and Malay labourers under European supervision stripped the island bare of vegetation before removing its overburden — the rich, phosphate sandstone cemented together by dissolved bird droppings. In 1893 Saville-Kent noted an island where "turtles and flocks of birds have become scarce ..." but curiously sought to blame the "disturbing influences of the lighthouse colony" and visits by "excursionists from the mainland", rather

Left: A Green Turtle lays her clutch of over 100 leathery, ping-pong ball sized eggs in the 50 to 60 centimetre-deep egg chamber that she has carefully excavated with her hind flippers. The eggs will take between 54 and 70 days to hatch. An average of 4,500 Green Turtles nest annually on the Capricornia cays, making the region one of world significance for turtle conservation.

A Black Noddy chick awaits the return of a parent with a feed of small fish, plankton or jelly fish. The first chicks hatch in December and take about 50 days to fledge. Both adults share parental duties.

Black Noddies nesting in the Pisonia grandis *forest on Heron Island.*

than the virtual decapitation of the island, for the decline in its wildlife.[9] By 1898 Lady Elliot's deposits were exhausted. North West Island was extensively worked at the turn of the century, with Fairfax and Lady Musgrave less heavily mined. Already greatly disturbed, Lady Elliot, Lady Musgrave and Fairfax were further degraded by goats left in the miners' stead as a living larder for shipwrecked mariners.

The harvest of turtles soon followed. A factory was established on North West Island in 1904 and operated intermittently, slaughtering Green Turtles (*Chelonia mydas*) for soup and extract oil until 1928. Heron Island was the site of a similar endeavour in the 1920s. A contemporary account stated that "To make one ton of extract it takes 440 turtles at 12 a day or 36 days. 100 cases of soup takes 228 turtles at 8 cases a day or 26 [*sic*] days."[10] The rate of plunder was obviously unsustainable and by the end of the decade the resource had been over-exploited. Water was scarce or tainted by guano and the Great Depression had taken hold.

Mercifully for the turtles, the factories closed, but not until 1968 did all species receive protection in Queensland.

As the intense exploitation of Capricornia's resources declined, scientific and recreational interest in the region grew. Turtle tagging and study commenced on Heron Island in 1929 and a research centre was established there in 1951. A second facility began on One Tree Island in the 1970s, contributing much to scientific understanding of the intricacies of coral reefs and cays. Tourism began in the 1930s when Heron Island's abandoned turtle factory was converted to a fishing camp, the genesis of today's resort. A string of island, then marine reserves followed. Finally, public outrage at the thought of oil prospecting on the reef precipitated the 1975 Great Barrier Reef Marine Park Act of the Federal Parliament. In 1981, Capricornia was the first part of a 350,000 square kilometre marine park to be declared and in the same year the Great Barrier Reef was inscribed on the World Heritage List.

Shifting Bulwarks

In the open ocean, seabirds eager for breeding space settle on emergent cays barely clear of the highest tide. Their droppings transfer nutrients from the sea to an infertile and barren landscape, giving a kick-start to colonising plants, the seeds of which are often carried to the islands by the birds themselves. Other hardy seeds are washed ashore and germinate after weeks at sea. Plants gradually contribute their own organic matter, promoting further growth. As the cay itself develops, calcium carbonate precipitates from percolating groundwater, cementing together slabs of sub-surface beach rock. Freshwater trapped in a sub-surface lens (a geological structure literally shaped in cross-sections like an optical lens) promotes forest growth. Size promotes stability.

But nothing remains static in Capricornia. The morning after our blustery visit to Erskine Island we went ashore on 40-hectare Masthead, second-largest of the Capricornia cays. The wind had abated considerably and was now blowing gently from the south-east. After landing, I made my usual circumambulation of a new island. The northern side of this seemingly immovable forested cay sported a dense stand of magnificent *Pisonia grandis* with a border of compact Octopus Bush (*Argusia argentea*), all backing a broad foredune and beach. Yet to the south, the forest edge was ragged and untidy. A jumble of bleached dead limbs and tottering branches sprawled over a narrow stretch of sand. The soil profile of the island was fully exposed as trees toppled, their roots bare and undercut by the wash of wave and tide.

The island was slowly being swept from under itself. Not that this large coral cay was in imminent danger of disappearing, but through the action of wind and water it was constantly shifting and reforming in its own image. As the southern flanks disappeared so the opposite side grew.

Further evidence of the dynamism of the cays can be gleaned from historical

A grove of Pandanus tectorius *on Lady Musgrave Island. Such salt-tolerant vegetation often rings the cays behind the beach and provides a wind-break for the central forest zone on the larger cays.*

Wedge-tailed Shearwaters occupy their dense nest colonies in Capricornia from October to April-May. Nesting occurs on all but North Reef and One Tree, with large colonies on Masthead and Heron islands.

Opposite: A climax forest of Pisonia grandis on Heron Island. Pisonia is a favoured but frequently deadly nesting choice for Black Noddies. The trees provide an abundance of stable nest sites and an ample supply of building material. In exchange, they are fertilised and seed is distributed. But the sticky seeds often become glued to the birds in such numbers that they are made flightless. No longer able to feed, they starve to death and so continue to fertilise the Pisonia.

comparison. One of the earliest descriptions of the flora of the cays was made by Jukes on Lady Musgrave Island in January 1843. He described the centre of the island as a ridge of coral rubble surrounding a small sandy plain, noting that "the encircling ridge was occupied by a belt of small trees, while on the plain grew only a short scrubby vegetation, a foot or two in height".[11] Jukes may have visited an island recovering from cyclonic damage for just 150 years later, Lady Musgrave supports a dense climax forest dominated by *Pisonia*. The flourishing forest appears even more remarkable when considered against the assaults of guano mining and the voracious mouths of up to 300 goats that ate everything that stood still until eventually shot out in 1971.

But while each cay moves, decays and is replenished, season after season, storm after storm, the wildlife of the region is provided with reliable breeding places on these shifting reef-top bulwarks.

Pisonias and Noddies

A pair of waterbirds patrolled the mirrored shallows at dusk on Lady Musgrave Island's lagoon. One was pure white, the other a slate-grey yet both were the same species: Eastern Reef Egrets (*Egretta sacra*) exhibiting the polymorphism for which they are well known. As the name implies, these statuesque birds are habitués of the reef-flat. Common on coastal waters throughout northern Australia, each bird defends its own feeding ground, hunting small fish and crustaceans at low tide. On the islands of Capricornia, they are a frequent sight rising from their favoured roosting trees, ever wary of human presence.

The white bird inched forward, then paused, hunched in an almost arthritic stance before firing its rapier-like bill at its target. Behind the beach more egrets were preparing for the night, taking roost as a loose congregation among the dense forest-edge of *Argusia* and Pandanus (*Pandanus tectorius*). Meanwhile, out at sea, clouds of Black Noddies (*Anous minutus*) drifted like smoke, dipping and snatching at shoals of small fish. Hundreds swept towards the island, their harsh cackling call filling the air. Black shapes swerved and fluttered out of the lengthening gloom, settling for the night in restless, jabbering groups throughout the *Pisonia* forest. Before dawn the flocks would again be feeding at sea or loafing in large "rafts" when wind and wave allowed.

Mating would soon begin. Birds had started to form pair-bonds — sitting close on branches while bill-sparing, or gathering on the ground, nodding and bowing to one another. Come November, a single egg is laid in a nest made of twigs and leaves cemented together with generous quantities of the birds' own excreta. Tiers of nests in their thousands, both new and used, festoon the *Pisonia* trees on several Capricornia cays.

Black Noddies breed in large numbers also on Heron, Masthead and North West islands. More than 155,000 pairs were estimated in the mid-1980s on North West alone. While Black Noddies breed on a few forested cays elsewhere on the Great Barrier Reef, Capricornia is their stronghold.

Collectively, the islands of Capricornia are the most important breeding ground for seabirds on the entire Great Barrier Reef. As many as one million Wedge-tailed Shearwater (*Puffinus pacificus*) pairs arrive at their breeding colonies in October each year. Favouring sand, the Shearwaters nest in burrows in large concentrations on North West, Masthead, Tryon, Heron and Lady Musgrave islands. Meanwhile, ground-nesters such as Crested, Black-naped, Bridled (*Sterna anaethetus*) and Roseate Terns (*S. dougallii*) breed in scattered, exposed colonies on bare sand and coral rubble.

From our anchorage beyond the reef crest, Tryon Island appeared almost to levitate atop its glistening kerosene-coloured lagoon. It was high tide and the reef-flat was awash. As we sped ashore in our inflatable, I noticed a dark boulder-like shape punctuating the island's neat white collar of sand. Storm debris I thought. Landing 50 metres down the beach, it was immediately clear that this was not some dreary piece of flotsam but a turtle. Large numbers of Green Turtles had arrived to mate in the Capricornia lagoons. They had migrated from distant feeding grounds as far afield as the waters of New Guinea, New South Wales or

Vanuatu. Capricornia is the focus of turtle breeding throughout the entire southern half of the Great Barrier Reef.

The turtle hauled up on the beach was an adult female. She had probably come ashore to snatch a break from a group of three or four amorous males. Even as she rested, they patrolled the shallows anxious for her to regain the water and resume mating. Each female may copulate with several males during the two-month mating season. Once concluded, the males return to their feeding grounds while the females remain in the waters close to their nesting beaches. They haul up on the sand cays to lay during January and February before returning to their home waters. Hatchlings emerge after 7–12 weeks. Their chances are slim. Less than one in 1,000 will survive the 40 years to sexual maturity and their turn to mate in the lagoons of their own conception.

The Ephemeral and the Immutable

Tamaru returned to anchor in the lee of North West Island. At 105 hectares, North West is the biggest cay in Capricornia and the second-largest on the entire Great Barrier Reef. Within the dense swathe of its *Pisonia* forest, the tents of 150 campers were almost completely hidden from view. It was low tide. Out on the reef-flat a string of people stooped in curiosity at the startling parade of life within its aquatic grottoes while a line of beached aluminium dinghies glinted in the dazzling sun. On the rising tide the boats would be off to favourite fishing spots along the drop-off from the reef edge.

While the flagrant abuse of Capricornia has long passed, a concentrated human presence alone places a strain on the intricacies of reef life. Unintentional damage from reef walkers' feet, overfishing, anchor damage and disturbance of ground-nesting birds and turtles are just some of the problems of a Great Barrier Reef under great pressure.

But perhaps the greatest challenge to these island gems does not lie within the ambit of those responsible for the stewardship of the reef. Increasing population pressures, coastal development and agriculture are all affecting the quality of reef waters. On a much larger scale, the challenge to the Great Barrier Reef is part of a great and looming global problem. A rapid "greenhouse"-induced climate change over the next 50–100 years is likely to be attended by an increase in rainfall, cyclone frequency and a rise in sea level. Under such circumstances the tenuous balance between the ephemeral and the immutable may be sorely tested.

1   True "barrier" reefs are found only north of Cairns, but the practice of referring to the whole reef system as the Great Barrier Reef is long established.

2   European exploration of the east coast of Australia may date back to the early sixteenth century and the voyage of the Portuguese navigator Christado de Mendonca. See K. G. McIntyre, *The Secret Discovery of Australia: Portuguese Venturers 200 Years Before Captain Cook*, Souvenir Press, Medindie, SA, 1977.

3   P. P. King, *Narrative of a Survey of the Intertropical and Western Coasts of Australia Performed Between the Years 1818 and 1822*, T. and W. Boone, London, 1827 (facsimile edn, Lib. Bd of SA, 1969), vol. 1, pp. 180, 352. Lady Elliot Island was discovered by Captain Thomas Stewart sailing on the merchantman *Lady Elliot* between Sydney and Batavia.

4   J. Beete Jukes, *Narrative of the Surveying Voyage of HMS Fly*, T. and W. Boone, London, 1847.

5   *Ibid.*

6   In June 1803, Matthew Flinders returned to Sydney in the leaky *Investigator* after abandoning his survey of the Australian coast. Unable to procure a suitable replacement in the colony, he set sail for England to plead his case. However, Flinders' plans were dashed when the *Porpoise*, on which he sailed as a passenger, together with its consort the *Cato*, struck reefs in the Coral Sea. Flinders went for help, returning to rescue all but four of his companions.

7   Early accounts of the Queensland industry refer to "bêche-de-mer", whereas accounts of Macassan involvement elsewhere refer to "trepang".

8   The principal investor in the Anglo-Australian Guano Company was W. L. Crowther of Hobart. Crowther owned an extensive whaling fleet and was one of the first Australians to land sealing gangs on Heard Island.

9   W. Saville-Kent, *The Great Barrier Reef of Australia; Its Products and Potentialities*, London, 1893 (facsimile edn, John Currey O'Neil, Melbourne, 1972), p. 103. A permanent lighthouse was built on Lady Elliot Island in 1873.

10 Correspondence relating to turtle factory on North West Island, quoted by C. Limpus, "Uncertain Land of Plenty", in H. J. Lavery (ed.), *Exploration North*, Richmond Hill Press, Melbourne, 1978, p. 221.

11 Beete Jukes, *op. cit.*.

CAPRICORN BUNKER GROUP

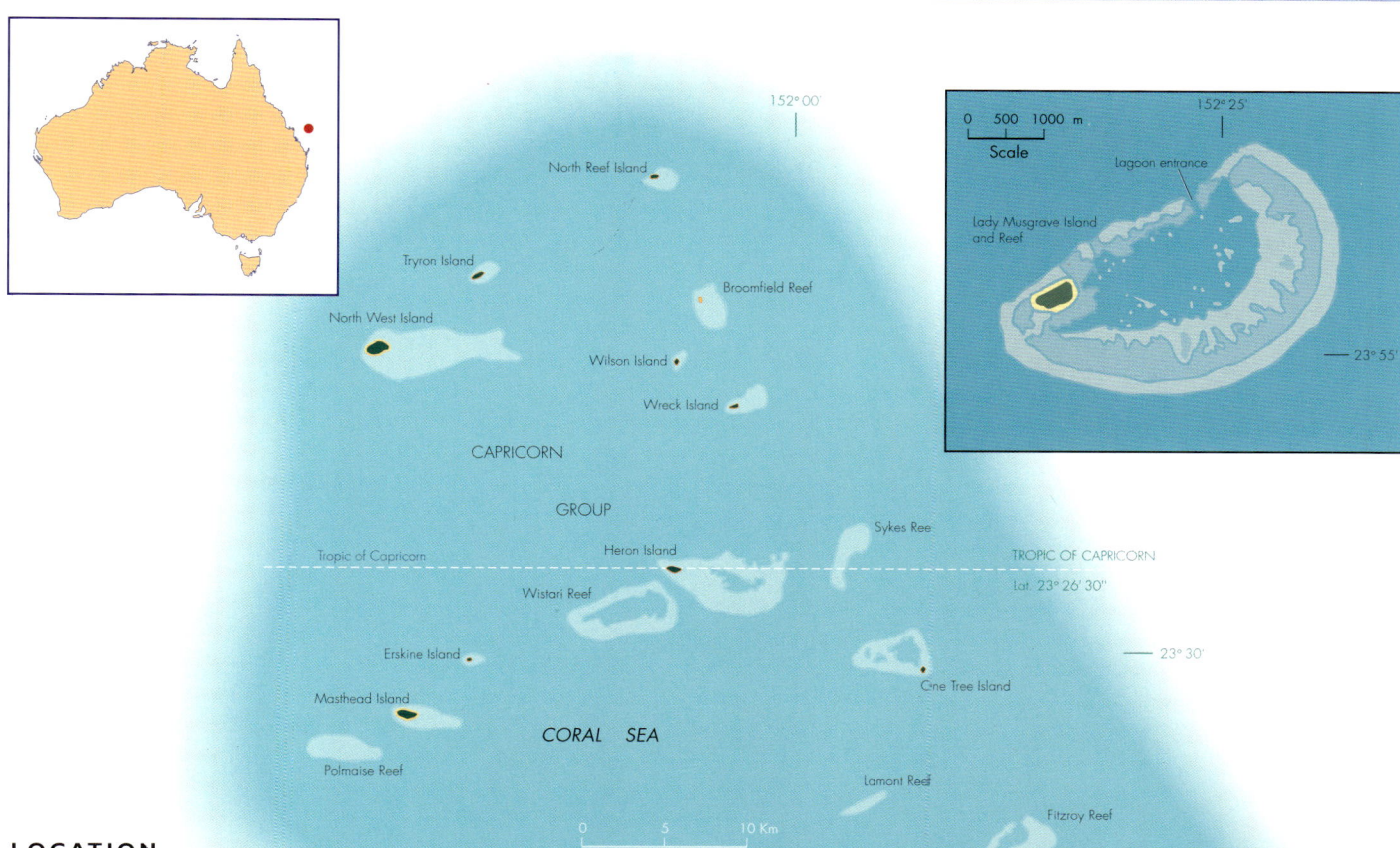

LOCATION

From North Reef Island (23°13' S, 151°57' E) to Lady Musgrave Island (23° 54' S, 152° 25' E) and Lady Elliot Island (24° 07' S, 152°43' E), the group straddles the Tropic of Capricorn, about 50 to 120 km east of the Queensland coast. 21 reefs, 13 supporting islands (9 in the Capricorn Group and 4 in the Bunker Group). The closest town is Gladstone.

AREA

North West Island: 105 ha
Lady Musgrave Island: 27 ha
North Reef Island: 1 ha

CLIMATE

Warm, subtropical/tropical

STATUS

The islands comprise the Capricornia Cays National Parks with the exception of Lady Elliot and North Reef islands which are Commonwealth property (reserved for lighthouses). Together with their surrounding reefs, the islands form part of the Mackay Capricorn Section of the World Heritage listed Great Barrier Reef Marine Park. Leasehold concessions on Heron, Wilson and Lady Elliot Islands. The Capricornia Cays National Parks are managed by Queensland's Department of Environment. The Great Barrier Reef Marine Park is jointly managed by the Department of Environment and the Great Barrier Reef Marine Park Authority (GBRMPA).

ACCESS

Air to Lady Elliot, helicopter to Heron and floatplane to Lady Musgrave Island. Regular catamaran services to Heron, Wilson and Lady Musgrave islands. One Tree Island and its surrounding reef is zoned for Scientific Research—access restricted to scientific researchers with permits. Hoskyn, Fairfax and Wreck islands are gazetted as National Park (Scientific) with access restricted above high water. Wreck Island's surrounding waters are also a Preservation Zone—access prohibited. Permitted activities vary on other islands, reefs and marine park waters through multiple-use zoning. Consult the Department of Environment or GBRMPA.

FACILITIES

Campgrounds on Lady Musgrave, Masthead, Tryon* and North West Islands.

Permits are required—consult the Department of Environment. Resorts with full amenities on Heron, Wilson and Lady Elliot Islands.
* Tryon Island now closed indefinitely to campers due to safety considerations from unstable trees.

A World Unto Itself (Mount Lidgbird from Mount Gower, Lord Howe Island)
Oil on canvas 122 x 107cm

HAVENS NEAR AND FAR

LORD HOWE ISLAND

CABBAGE TREE ISLAND

SPECTACLE ISLAND

Separated in time and space, each of these islands exists as a world unto itself. From two overlooked orphans of the mainland to one of the world's greatest evolutionary exemplars, this trio highlights the opportunities and perils of island life: imperatives that have global significance in this dire age of extinctions.

LORD HOWE ISLAND

Oceanic Solitude

The locals call it the Getting Up Place — which is a rather matter-of-fact name that says nothing of the heart-pounding thrill of hanging onto the side of a mountain in the sea, peering down through a hole in the mist to the lagoon far below. A streak of turquoise, lace-edged with surf, drifted in and out of view as cloud enwrapped our lofty vantage point.

Out across some weathered slabs, and then almost straight up was the only way ahead. A 20-metre band of the climb's steepest and most exposed rocks had to be negotiated before the track at last began to flatten. Reaching the summit plateau of Mount Gower, we were more than 800 enervating but exhilarating metres above the Tasman Sea on the edge of an undulating mesa, girt by cliffs on all flanks. The narrow ridge we had just climbed was the only way up and down. Mount Gower's summit covers more than 27 hectares and reaches an ultimate height of 875 metres near its southern end — Lord Howe Island's highest point.

Such is the island's isolation that Mount Gower towers above great expanses of empty ocean. The closest landfall is 570 kilometres to the west, at Port Macquarie, and the nearest island neighbours are Norfolk, 800 kilometres further on into the Pacific, and New Caledonia, 1,250 kilometres to the north-east. Solitude as wide as an ocean has done much to fashion both the natural and human history of Lord Howe Island.

Violent Beginnings

Mount Gower is not so much a peak in the "classical" sense as a brutish massif of forest-clad basalt. It sits in close consort at the southern end of the island with the slightly lower but no less imposing 777-metre Mount Lidgbird. Together, the mountains dominate more than just the physical landscape. They are a constant, brooding reminder of the dramatic origins of the place.

Between the late Cretaceous and early Tertiary Periods, 60–80 million years ago, a 2,000-kilometre segment of continental crust known as the Lord Howe Island Rise was torn from Australia, causing the intervening ocean floor to spread and sag. This seabed bottomed out more than 4,000 metres below current surface levels as the Tasman Basin formed. Geological epochs passed and the Australian tectonic plate inched northwards, carrying the Lord Howe Island Rise with it. Beneath the relatively shallow boundary between the Tasman Basin and the Lord Howe Rise, a stationary "hot-spot" in the earth's mantle extruded a 1,000 kilometre-long chain of submarine mountains as the plate ground past. The first of these seamounts formed about 23 million years ago. As each grew in turn, it was shunted past its magma source and the process began afresh.

The closest seamounts to Lord Howe Island are those beneath Elizabeth and Middleton Reefs, 140 kilometres to the north and more than 3 million years older. Lord Howe and its surrounding islets sit atop an enormous truncated pedestal at the southern end of the seamount chain.[1] Comparatively young, the island group emerged from the sea about 6.9 million years ago or some 16 million years after its underlying seamount began to grow.

Life started violently for Lord Howe. The island broke the surface as an explosive mix of ash, steam, lava "bombs" and rock hurled from vents in the vicinity of Malabar and the Admiralty Group. Successive eruptions over the next 500,000 years created a broad shield volcano, twice the island's present height and with an ultimate breadth of 30 kilometres.

Volcanic activity finally shifted to the vicinity of Mount Lidgbird where a large collapsing caldera formed within the summit of the volcano. Ten square kilometres in area and nearly a kilometre deep, the caldera rapidly filled with basaltic lavas in flows up to 30 metres thick. This final spectacular phase of volcanism

Opposite: From North Bay looking across the protected waters of the Lagoon to Mount Lidgbird (left) and Mount Gower at the southern end of Lord Howe Island. Rising to nearly 900 metres straight out of the Tasman Sea, the two mountains are often shrouded in cloud. This results in a distinct microclimate across their higher flanks.

ended about 6.4 million years ago. Since that time erosion has reduced the island to a mere fraction of its former self. Less than three per cent of the island's former bulk is left today. Lord Howe is now only 11 kilometres long and less then three across at its widest point.

Twenty-three kilometres to the south-east, and born of the same seamount, one of the tallest and most isolated sea stacks on earth knifes skyward. Balls Pyramid is the 551 metre-high remnant of a second and smaller shield volcano thought to have been active at about the same time as Lord Howe.

From Supply to Demand

From a small clearing on the edge of the summit plateau, Mount Lidgbird loomed large just a kilometre or so across the sweep of Big Hill Saddle. This was not the great bluff seen from the settled parts of the island but a spectacular peak, the same mountain in a totally different guise. Its sides fell away in all directions in a series of precipitous slopes and rock walls the height of 30-storey buildings. Layer over distinct layer of grey basalt had registered each injection of lava into the long-vanished caldera.

While more than six million years of erosion has radically reduced the size and shape of the island, little of its superficial appearance has changed since the first known human encounter with Lord Howe. As one of the most isolated places in the Pacific, it eluded both the great Polynesian migrations and European voyages of exploration.

Barely two weeks after the arrival of the First Fleet at Sydney Cove in January 1788, Lieutenant Henry Lidgbird Ball in command of HMS *Supply* headed north-east for Norfolk Island. Found in 1774 by James Cook, Norfolk was considered to be of great strategic importance. A convict settlement was to be established there without delay.[2]

En route, Ball sighted a spectacular cluster of mountains rising from the sea. Two months later on the return passage to Sydney, he landed, claimed possession for the Crown, naming several features for himself and naval colleagues. The largest member of the group was named in honour of Richard, Earl Howe, First Lord of the Admiralty.[3]

And so ended the island's innocence. It immediately became a provisioning and watering stop for ships sailing between Norfolk Island and Sydney. Settlement eventually followed. In June 1834 three men, their wives and two Maori "boys" were landed from a passing whaling ship, the *Caroline*. They cleared parts of the lowlands, built simple huts from locally won timber and palm thatch, and began farming everything from carrots to sugar cane. Liberated by early visitors, goats and pigs prospered and provided a ready supply of meat. The islanders traded their produce with any passing ships but most of their business came from whalers active during the middle decades of the century in the Tasman Sea and mid-Pacific.

A few more settlers arrived and at the height of the whaling era, 60–70 ships called at Lord Howe each year. However, by the 1870s whaling was in decline. Trading routes changed with the advent of steam power. Fewer ships stopped and the welfare of the islanders looked to be in peril. But an unlikely quarter — the fashionable drawing rooms and palm courts of the Victorian and Edwardian eras — provided a welcome solution. Distinctive among the island's rainforest vegetation are four endemic species of palm. One in particular, the so-called Kentia or Thatch Palm (*Howea forsteriana*) became a favourite parlour specimen and is still one of the world's most popular indoor plants.[4] Seed was first exported from the island in 1890. Despite setbacks due to war in Europe and the introduction of rats from a grounded freighter in 1918, seed collection and palm propagation has remained an important contributor to the local economy.

A regular shipping service began operating to Lord Howe in 1893. Holiday visitors soon arrived, at first on freighters, then after 1932 on a dedicated run from Sydney. In the years following World War II, the island gained in popularity and prosperity, with thousands of visitors arriving each year by flying boat and later by light aircraft. An airstrip was built across the island in 1974 and now handles frequent traffic from Sydney, Brisbane and Port Macquarie.

Lord Howe Island settlers in front of a barn thatched with palm fronds. The photograph dates from the first pictorial record of the island made in 1880.
Government Printing Office Collection. Courtesy of State Library of NSW

Lord Howe remains as popular as ever. There are currently about 300 permanent residents. Tourists visit year round, attracted by the island's natural wonders, even subtropical climate and relaxed pace. But Lord Howe is no garish resort destination: its charm and beauty are safeguarded through limits on visitor numbers. Fewer than 400 tourist beds are distributed through 17 guesthouses and lodges, quietly secreted among the lowland forests.

The Mountains of Providence

A strong, moist breeze had been blowing from the south-east all morning. Pushing up and over the mountains, it cooled and condensed as mist and cloud, shrouding the tops in a spectral veil. Silently, Mount Lidgbird vanished and the southern end of the island withdrew into its own isolated world.

Away from the dissolving view, something stirred in the dense vegetation. An inquisitive dark-grey bird waddled from a well-concealed burrow into the daylight. It was a young Providence Petrel or Big Hill Muttonbird (*Pterodroma solandri*) that appeared to have been abandoned by its parents. It being mid-December, the rest of the colony was on the wing somewhere in the distant Pacific.

About 20,000 pairs of Providence Petrels return to their colonies in the southern

mountains in March each year and breed through the winter months. Chicks hatch in July, with both fledglings and adults quitting the island in November to roam the Western Pacific as far north as Japan and Hawaii. The bedraggled individual had failed to thrive and was doomed, its demise an inevitable part of the natural process.

The circumstances that inspired the naming of the Providence Petrel are indicative of the fate of many of the original inhabitants of the primordial Lord Howe and its distant neighbour, Norfolk Island. It is a name filled with irony, considering the fate of the two islands' unique wildlife upon the arrival of humans.

In the grim days following the settlement of Norfolk Island, food was often critically short. For several years all that forestalled starvation was the indigenous wildlife. Enormous numbers of seabirds were killed. The most common species taken came to be known as the Bird of Providence. In just three months during 1790, more than 170,000 birds were slaughtered and by 1800 it was extinct on Norfolk Island.[5]

The Providence Petrel shows no fear and can even be called down from the air, such is its sense of curiosity. This primordial, almost innocent trust was common to a number of species at the time of the first human visits. As an isolated and undisturbed oceanic island, Lord Howe had been the setting for the evolution of a rich endemic fauna — from flies, spiders and snails to birds.

"... the golden age ..."

In such a benign environment with few predators, both land and seabirds thrived. With no reason to flee, some landbirds gradually gave up flying in an evolutionary adaptation common to islands; their energies were more profitably spent exploiting the forest floor.

Arthur Bowes Smyth, surgeon on the *Lady Penrhyn*, one of four First Fleet ships to call at Lord Howe in May 1788 gave a description of the island's birdlife, at times lyrical and yet depressingly portentous of the ultimate fate of several bird species:

"When I was in the woods amongst the birds I cd. not help picturing to myself the Golden Age as described by Ovid to see the Fowls or Coots some white, some blue & white, others all blue wt. large red bills & a patch of red on the top of their heads, & the Boobies in thousands, together wt. a curious brown bird abt. the size of a Land Reel [Rail] in England walking totally fearless & unconcern'd in all part around us, so we had nothing more to do than to stand still a minute or two & knock down as many as we pleas'd wt. a short stick. The Pigeons were also as tame...& wd. sit upon the branches of trees till you might go & take them off with your hand..."[6]

In 1788, the entire suite of indigenous landbirds is thought to have comprised 15 species. Thirteen were endemic to the island. Nine of these are now extinct. The spectacular White Gallinule (*Notornis alba*) and the large and abundant White-throated Pigeon (*Columba vitiensis godmanae*) described by Bowes Smyth, were early casualties. Both were considered good eating. The White Gallinule was probably extinct by the time of settlement and the White-throated Pigeon, an endemic subspecies of a bird widespread in the Pacific, was hunted out by the 1850s. The island sported a distinct subspecies of the Red-fronted Parrot (*Cyanoramphus novaezelandiae subflavescens*) which unfortunately took a liking to the settlers' crops and was exterminated by the 1870s.[7] Five small endemic landbirds, the Vinous-tinted Blackbird (*Turdus xanthopus vinitinctus*), Lord Howe Fantail (*Rhipidura cervina*), Lord Howe Warbler (*Gerygone insularis*), Robust White-eye (*Zosterops strenua*) and the Lord Howe Starling (*Aplonis fuscus hullianus*) all became extinct shortly after the arrival of rats. In 1940, the scale of the disaster moved ornithologist K. A. Hindwood to liken the loss of so many of Lord Howe's landbirds to the great extinctions that occurred on Madagascar and Mauritius.

Only two of the island's small forest birds, the Lord Howe Golden Whistler (*Pachycephala*

Small groups of dazzling White or Norfolk Terns (Gygis alba) frequent the lowlands during summer. Not bothering to build a nest of any kind, they lay a single egg in shallow notches or dents on bare branches. Consequently, eggs and chicks are extremely vulnerable in high winds or storms. All birds leave the island on the first winter gales, not returning until September.

A Red-tailed Tropicbird nesting in December at the Goat House on the flanks of Mount Lidgbird. Ranging across the lower latitudes of the Western Pacific and Indian Oceans, the Tropicbird's largest breeding concentration is thought to be on Lord Howe Island. In the summer breeding months they are particularly common around the mountains and the northern sea cliffs where spectacular aerobatic and courtship displays occur. In winder most birds quit the island and roam the open ocean.

pectoralis contempta) and the Lord Howe White-eye (*Zosterops tephropleura*) have survived to the present day. The White-eye remains the most common landbird, despite the rats and the deliberate introduction of the similar Australian Silver-eye (*Z. lateralis*).

Back from the Edge

Settling for lunch within the enchanted world of Mount Gower's forest, we were soon joined by the island's best known celebrity, the flightless Lord Howe Woodhen (*Tricholimnas sylvestris*). It was as fearless as the early accounts claimed. A single bird, then another appeared. With long curved bills, they scratched in the leaf-litter and humus close to my feet. Although probing for grubs, larvae, molluscs and insects, they were not adverse to quickly cleaning up any spilt crumb.

Unique to Lord Howe, until recently the Woodhen was one of the world's most critically endangered birds. Like the Gallinule and White-throated Pigeon, it was considered fine fare by early visitors and settlers. With the added burden of predation and disturbance of its habitat by feral cats, goats and pigs, Woodhen numbers had dropped to pathetically low levels by the 1970s. Mount Gower's summit

became their last stronghold. In 1971, an extensive search revealed that perhaps just 26 individuals remained.[8] A captive breeding program was begun in 1980 and during the following three years 92 chicks were artificially incubated and raised on the island. Pigs, as the greatest enemy, were vigorously pursued and successfully exterminated. As early as 1981, young birds were progressively released back to the wild. Although the Woodhen must still be regarded as vulnerable, the program was an outstanding success and the birds are now found in many parts of the island. The present population is thought to exceed 200.

It was a joy to see these trusting, inquisitive birds, strutting about their forest floor home and not as the lifeless museum specimens they nearly became. The cool heights of Mount Gower would be tragically empty without them.

Palms in the Clouds

Lord Howe's southern mountain forests are a setting deserved of such rare animals, for within just a few hectares, an array of plants has evolved that are found nowhere else on Earth.

During the final stages of our climb, the air had grown noticeably colder — temperatures on Mount Gower are 6–8° Celsius lower than at sea level. It is also much wetter in the mountains, with the frequent mantle of orographic cloud invariably bringing rain. Due to the extreme isolation and prevailing microclimate, the southern mountain tops and upper slopes support a high proportion of the island's endemic flora. Cool air and abundant moisture have conspired to create a unique, miniature world in both scale and extent.

A densely woven canopy less than five metres off the ground enclosed a realm of fanciful spangles and gossamer drapery. From orchids and filmy ferns sprouting among hirsute swathes of moss to arching tree ferns — green was everywhere in countless dripping hues. Despite being near the highest point for over 500 kilometres, I could see nothing of the world outside. The only way to appreciate the extent of the plateau forest was to climb a tree. Selecting a stout specimen, I struggled upwards through tightly entwined limbs. The tree's unyielding crown obscured much of the view but I was able to glimpse a small stretch of unbroken canopy disappearing into the mist.

Emerging from this undulating aerial carpet were several of the distinctive plants restricted to the southern mountains; *Dracophyllum fitzgeraldii*, the world's largest member of the

Epacridaceae or heath family, endemic tree ferns (*Cyathea* spp.) and the Pumpkin Bush (*Olearia mooneyi*). Blooming in early summer with dense clusters of white daisy-like flowers, the Pumpkin Bush is one of the common trees on the summit of Mount Gower. Also there were the Big Mountain Palm (*Hedyscepe canterburyana*), and the striking orange-flowered Pumpkin Tree (*Negria rhabdothamnoides*). Both common at elevations above 500 metres, the Big Mountain Palm and Pumpkin Tree are each the sole members of endemic genera.

The ancestry of Lord Howe's flora is mixed. Its dense and luxuriant forests are more strongly allied to sub-tropical and temperate rainforest than the sclerophyllous, eucalypt-dominated regions of Eastern Australia. Apart from five genera found nowhere else, the island has close botanical ties with Australia, New Caledonia, New Zealand and to a lesser extent Norfolk Island.[9]

This diverse heritage gives just an inkling of the distribution of land across an ancient Tasman Sea. At times of advanced worldwide glaciation, sea levels would have possibly been more than 100 metres lower than they are today, leaving seamounts exposed as island chains at the edge of the Lord Howe Island Rise

*The Lord Howe Island Currawong (*Strepera graculina crissalis*), an endemic subspecies. The Currawong and the now extinct Lord Howe Island Boobook (*Ninox novaeseelandiae albaria*) were the only predators on Lord Howe prior to settlement. The introduction of far more effective predators such as rats, Tasmanian Masked Owls (*Tyto novaehollandiae*) and humans made trusting, flightless landbirds extremely vulnerable to disturbance.*

*A much-banded Lord Howe Island Woodhen foraging on Mount Gower's summit plateau. The Woodhen has most likely been on the island for thousands of years and may be related to the Banded Landrail (*Gallirullus philippensis*), a common Pacific island coloniser. However, it is the sole member of its genus since its only close relative, a rail from New Caledonia, is now thought to be extinct.*

*Above: Sallywood (*Lagunaria patersonia*), a common tree of the lowlands, is also found on the Queensland coast and on Norfolk Island.*

*Opposite: Of the 219 native vascular plants found on Lord Howe, 74 are endemic. More than half of the 25 recognised plant associations are dominated by endemics. This richness and concentration of unique species is nowhere more apparent than in the gnarled mossy forests atop Mounts Gower and Lidgbird. The two palm species found at altitude, the Big Mountain Palm and the Moorei or Little Mountain Palm (*Lepidorrhachis mooreana*), are each sole members of individual endemic genera.*

and along the length of the Tasman Basin. Birds would have easily negotiated these short divides. As early colonists, they probably arrived in the course of evolving migratory patterns or perhaps as unwilling vagrants swept up in storms. Seeds and spores may have arrived on the wind, but must also have reached the island as casual baggage on feathers and feet, or safely stowed within avian alimentary tracts.

The independently evolved vegetation of Lord Howe and its diverse origins were but two factors leading to its inscription on the World Heritage List in 1982. Despite more than 150 years of human settlement, lowland clearing and the introduction of 160 alien plant species, most of Lord Howe's forest cover remains intact and virtually undisturbed.

The afternoon was well advanced when we left Mount Gower. Cloud still caressed the summit of Mount Lidgbird but was high enough to allow a glimpse of tiny Muttonbird Island riding a sullen sea, far below. Two kilometres off the east coast of the main island, Muttonbird is one of several craggy outliers that are the strongholds of Lord Howe's seabirds.

Above and Below Among the Admiralties

On my last full day on the island, the south-easterly that blew for much of the previous week had thankfully dropped below 15 knots. We dashed from behind the protection of Mount Eliza's tottering cliffs, across one kilometre of open water to Roach Island in the Admiralty Group. Malabar, a headland of 200 metre-high basalt, loomed over our starboard side. Scarfaced with incisions from the numerous dykes squeezed through their fabric more than six million years past, these pockmarked cliffs are today home to thousands of breeding seabirds. Engrossed in aerobatic courtship display, great numbers of Red-tailed Tropicbirds (*Phaethon rubricauda*), brilliant white against the rockface, soared, propped and dived with balletic grace. Among the arabesques of the Tropicbirds, thousands of Sooty Terns (*Sterna fuscata*), the most numerous seabirds nesting on Lord Howe

Island, swarmed around the cliffs like bees near a hive.

Unlike their terrestrial cousins, the indigenous seabirds of Lord Howe did not suffer as badly with the advent of humans, although the extent of several species is now significantly diminished. The White-bellied Storm Petrel (*Fregetta grallaria*), for instance, once nested among rocks on the mountain slopes but is now confined to the offshore islets. There has probably been only one extinction since 1788. The large and mysterious Tasman Booby (*Sula tasmani*), known only from sketchy early descriptions and fossil remains, is thought to have occurred on both Lord Howe and Norfolk Islands prior to human contact. Nesting on dunes behind the lagoon, it must have been easy prey to those stepping ashore and was one of the first species sent to extinction.

The persistent south-easterlies made landing in the lee of Roach Island quite easy. But moving about on its steep slopes was not so simple. Rather than cause pandemonium among thousands of birds, I decided to savour the spectacle from the sidelines. The steep western faces of the island were clad in tussocks of *Poa poiformis* that barely concealed whole hillsides riddled with the burrows of Wedge-tailed Shearwaters (*Puffinus pacificus*). Preferring the offshore islets, about 30,000 pairs nest on the Admiralties and other outliers. Smaller colonies occur on main island promontories. Of the other two *Puffinus* species breeding on the Lord Howe Group, the Fleshy-footed Shearwater (*P. carneipes*) congregates in the sandy coastal forests of the main island's north-eastern end but the Little Shearwater (*P. assimilis*), like the Wedge-tailed, also burrows on Roach Island. Approximately 4,000 Little Shearwater pairs nest about Lord Howe, nearly all on Roach. It is the largest breeding site in the Australian region for this widely distributed sub-tropical species.

A dive boat was anchored near Roach Island. One by one divers surfaced and awkwardly clambered aboard. Without doubt they had just witnessed a small part of one of the Pacific's great marine environments. Later that day I joined them in exploring Lord Howe's aquatic menagerie, again near the Admiralties.

On the western side of the main island, the world's most southerly coral reef flanks the coast from North Head, to below Mount Lidgbird. Here corals flourish more than 1,000 kilometres south-east of the Great Barrier Reef. Thanks to a spasmodic swipe from the warm tail of the southward-flowing Eastern Australian Current, tropical larvae from Queensland and New Caledonia mingle with those of more temperate origins. As a result, an array of 83 hard corals, both tropical and subtropical, occur in Lord Howe waters. This is significantly less than the Great Barrier Reef but is nonetheless remarkable considering the island's isolation, small reef area and southerly latitude.[10] Species turnover is high. Formative tropical organisms arrive in the drift but do not necessarily establish viable local populations. Others move in and take their place, perpetuating change in a dynamic marine environment.

An Apparition of Paradise

We descended to the top of a turret of rock studded with lurid coloured sponges, sea feathers and tufts of algae among encrusted coral plates. Down into a narrow alley, spectacular gorgonian fans sprouted from its walls, their delicate filigree snaring microscopic food from the passing stream. Such a profusion of life — all founded on black basalt. Rather than the living being turned to stone, the stone had been engulfed by life. Several Butterfly Cod or Firefish (*Pterois volitans*) lurked in the shadows, their beautiful but menacing spines quivering a warning not to venture too close. The alley in the seafloor opened onto an expanse of sand running off into the depths. In the distance great numbers of pelagic fish cruised by. Among the stream were sleek King Fish (*Seriola lalandii*) and Spangled Emperors (*Lethrinus nebulosus*), glinting as they changed course, countless fish moving as one.

A Green Turtle (*Chelonia mydas*) flew out of the gloom, swerved, then vanished as quickly as it had appeared. Although not an uncommon sight in the island's inshore waters, turtles used to be far more abundant. Early visitors repeatedly took away from the lagoon "as many turtles as they could stow".[11] This apparition was a reminder of what Lord Howe once was and yet still persists — even as a shadow of its former self.

Tiny Lord Howe was the last significant Pacific island to be discovered. That it was found within the span of recorded history allows for poignant comparison across the entire region, between the primordial Pacific and what remains. Even so, the original plenty of this ocean paradise can now only be dreamed of. Valued against the loss of so much, Lord Howe Island must surely remain one of the world's most extraordinary places.

Opposite: Looking south from Roach Island through skies filled with Sooty Terns to Lord Howe. Twenty-three kilometres to the south-east, Balls Pyramid soars skyward 551 metres. Roach, largest of the Admiralty Islands, is home to the group's major concentration of seabirds. As well as Sooty Terns, the largest of Lord Howe's birds, the Masked Booby (Sula dactylatra) finds Roach Island's open grassy slopes ideal for nesting. The Lord Howe Group is its most southerly breeding colony.

Sooty Terns on the summit of Mount Eliza at the northern end of Lord Howe Island. They are the most numerous seabirds to breed in the Lord Howe Group, with large colonies occuring on the open ground of headlands and offshore islets. Sooty Terns breed and care for a single chick through spring and summer, then disperse, roaming the mid-Pacific during winter before returning to Lord Howe in July.

Brown Anemone or Clown Fish (Amphiprion akindynos) brooding its eggs within the protective carpet of its host anemone. Several species are found at Lord Howe Island with McCulloch's Anemone Fish (Amphiprion maccullochii) a particularly common sight among Lord Howe Island's reef and coral communities. In exchange for sanctuary within the anemone's stinging tentacles, the anemone fish does the "housekeeping" and deters predators.
Ian Shaw

1   The submerged platform on which the island sits shows the possible limits of its original size, centred around the 50-metre isobath. Out to the west the ocean floor rapidly plunges to a depth of over 2,000 metres.

2   Lieutenant (later Governor) Phillip Gidley King was dispatched by Governor Philip in the *Supply* under Ball's command with the urgent intent of establishing a garrison and penal settlement on Norfolk Island. The prize was the magnificent stands of Norfolk Island Pine (*Araucaria heterophylla*) and abundant flax plants — the raw materials for spars, sails and rigging, and the maintenance of naval supremacy. See G. Blainey, *The Tyranny of Distance*, Sun Books, Melbourne, 1966.

3   What appeared to be the highest mountain, Ball named for himself, along with the conspicuous Balls Pyramid. He honoured a second naval superior, naming Mount Gower for John Leveson Gower, a junior Lord of the Admiralty.

4   When specimens were first collected in 1870, the botanist Baron Ferdinand von Mueller wrongly ascribed *H. forsteriana* to one genus *Kentia*. *H. forsteriana* and *H. belmoreana* were placed in their own genus in 1877 but the common name Kentia persisted.

5   P. G. Fildon and R. J. Ryan (eds), *The Journal and Letters of Lt. Ralph Clark 1787–1792*, Australian Documents Library, Sydney, 1981, p. 193.

     Until recently the Providence Petrel was considered a de facto endemic species on Lord Howe Island but in 1985 a small breeding colony was found on Philip Island, an outlier of Norfolk Island.

6   P. G. Fildon and R. J. Ryan (eds), *The Journal of Arthur Bowes Smyth: Surgeon*, Lady Penrhyn *1787–1789*, Australian Documents Library, Sydney, 1979, p. 84.

7   Two closely related forms have suffered similar fates: the Norfolk Island Parakeet (*C. n. cookii*) is critically endangered and the Macquarie Island Parakeet (*C. n. erythrotis*) is also extinct.

8   H. J. de S. Disney, 'Survey of the Woodhen' in H. F. Recher and S. S. Clerk (eds), *Environmental Survey of Lord Howe Island, A report to the Lord Howe Island Board*, Australian Museum, Sydney, 1974, pp. 73–6.

9   Lord Howe Island shares 120 plant genera with Australia, 102 with New Caledonia, 75 with New Zealand and 66 with Norfolk Island.

10 The Great Barrier Reef supports 356 coral species. Like Lord Howe, the Solitary Islands (29° 55' S to 30° 14' S), close to the New South Wales coast near Coffs Harbour, support a rich mix of tropical, sub-tropical and temperate marine species. Seventy-seven hard corals of tropical origin have been recorded there.

11 Thomas Gilbert, Captain of the *Charlotte*, 1788; quoted in I. Hutton, *Birds of Lord Howe Island, Past and Present*, Ian Hutton, Coffs Harbour, 1990, p. 13.

LORD HOWE ISLAND

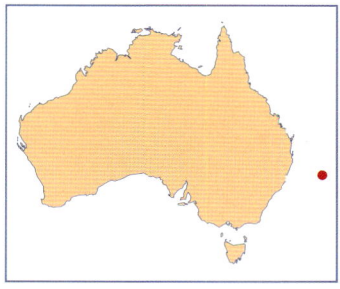

LOCATION

31° 33' S, 159° 05' E. 570 km east of the NSW coast, 702 km north-east of Sydney. Closest town is Port Macquarie.

AREA

Lord Howe Island: 1,455 ha
Roach Island: 16 ha

CLIMATE

Equable, sub-tropical. High rainfall.

STATUS

75% of the main island and all offshore islets and rocks are contained within the Lord Howe Island Permanent Park Preserve, inscribed on the World Heritage List. The Lord Howe Island Board is responsible for the Permanent Park Preserve and local government.

ACCESS

Regular commercial and charter flights from Sydney, Brisbane, Coolangatta, Coffs Harbour and Port Macquarie. Yacht access to secure anchorage within the lagoon. Access to the summit of Mt Gower allowed only in the company of an approved guide.

FACILITIES

17 guest houses and lodges offering self-contained units. Comprehensive visitor amenities.

Tenth of June Island
Admiralty
Islands
Roach Island
Noddy Island
South Island
159° 03'
159° 05'
Sugarloaf Island
Soldiers Cap
0   1   2 Km
31° 31'
Mount Eliza (147)
Kims Lookout
Malabar Hill
Neds Beach
North Beach
Old Settlement Beach
North Head
Middle Beach
TASMAN SEA
Clear Place Point
The Clear Place
The Lagoon
Lagoon Beach
Coral Reef
Blackburn Island
Blinky Point
Blinky Beach
Sail Rock
Airstrip
BLINKENTHORPE BAY
Mutton Bird Island
31° 33'
Mutton Bird Point
Intermediate Hill (250)
Rocky Run
Boat Harbour
Boat Harbour Point
Goat House Cave
Far Flats
Mount Lidgbird (777)
Red Point
Erskine Valley
The Big Saddle
Worlds End
31° 35'
The Little Slope
Mount Gower (875)
The Big Slope
King Point
Gower Island

CABBAGE TREE ISLAND

Island Gem

The Gould's Petrel (*Pterodroma leucoptera*) is one of the world's rarest sea birds. First described in 1844 by the naturalist John Gould, it is distinguished from other petrels by a sooty black head, nape and tail and brilliant white underparts. Its population is perilously small, with only around 300 breeding pairs. It seems the birds spend most of their life at sea, roaming parts of the Tasman and the Pacific. But much about the Gould's Petrel remains a mystery. One aspect of its life is, however, well known: each summer adult birds return to the world's only known nesting site for these ocean wanderers, a small gem of an island named Cabbage Tree.

Since 1992 the nesting Gould's Petrels have been the focus of a scientific study conducted by the New South Wales National Parks and Wildlife Service, with the support of the Australian Nature Conservation Agency (ANCA). Its aim is to determine the status of the petrel population and the factors affecting their breeding success and survival. From October to May the nest sites are checked morning and night to record patterns of occupancy, incubation and fledging.

Although located one and half kilometres from the mainland, Cabbage Tree Island seems much further removed. Within minutes of leaving the marina at Nelson Bay we rounded Yacaaba Headland and the clamour of Port Stephens was shielded from sight. Ahead lay the densely wooded slopes of Cabbage Tree. The island is only a kilometre long but the steepness of the terrain made it appear larger. From a rocky landfall at a small cove, we went up into steep gully. Only a few metres from the rocky shore we were engulfed by luxuriant vegetation.

Basecamp for the study program is perched near the foot of a scree slope. A fibreglass 'Apple' dome, the kind normally deployed as portable field stations in the Antarctic, provides all-weather accommodation. Poised on a raised platform of boulders, this green igloo nestles in the forest fringe alongside an awning-covered kitchen and dining area. Run-off from the awning feeds a water tank and solar panels provide power for lighting and water pumps. It is a cosy human nest, reflecting the kind of care and ingenuity being brought to bear on the problems of the beleaguered petrels.

Forest Nests

Littoral rainforest covers nearly 70 per cent of Cabbage Tree Island's 23 hectares. Prominent species include Native Olive (*Olea paniculata*), Guioa (*Guioa semiglauca*) and Tuckeroo (*Cupaniopsis anacardioides*). The island has six species of fig, the most ubiquitous being the Whalebone Tree (*Strebulus brunonianus*). Emergent trees in the canopy include large specimens of Deciduous Fig (*Ficus superba* var. *hennaeana*) and Strangler Fig (*Ficus watkinsiana*).[1]

The boulder scree rising above the camp is one of two parallel gullies spanning the western slope of the island. Cabbage Tree Palms (*Livistona australis*) are the most distinctive feature in these tracts of forest, with some specimens rising more than 10 metres tall.

While the forest canopy is lush, at ground level there is little in the way of an understorey. As we climbed up the moss and lichen-covered boulders of the South Gully, we noticed wire stakes dotted around the scree, numbered and flagged, marking known and occupied petrel nest sites. The Gould's Petrel typically lays its

Accomplished in the air but ungainly on land, Gould's Petrels often take time to recuperate after their perilous twilight landings through the forest canopy.

Opposite: A wave bursting through a kelp-lined cleft on the northern tip of Cabbage Tree Island. Cabbage Tree is one of the northernmost strongholds of the Little or Fairy Penguin, which makes landfall on these rocky shores to breed.

Above: Among the fig roots and ferns, the fallen fronds of the Cabbage Palm give the forest floor a striking appearance, and can also serve as rudimentary shelter for nesting petrels.

Opposite: Dense forest flourishes on the sheltered western face of Cabbage Tree Island, where two scree-filled gullies provide breeding habitat for the Gould's Petrel.

single egg in small rocky crevices. Minimal nesting material is used, though Cabbage Palm fronds littering the ground play an important role in protecting birds from predators and the elements. Other makeshift sites include the hollow trunks of fallen palms and nooks under the buttressed roots of large trees. Recently, however, the petrels have been presented with another nesting option.

Far from being passive observers, those involved with the research program are committed to helping this species pull back from the brink of extinction. Surveys carried out at the start of the program indicated that the population had declined by more than 30 per cent since a 1970 survey. And the success rate for breeding was only around 20 per cent.[2]

The Gould's Petrels are not especially proficient as parents. Their *ad hoc* nests often lack shelter, and eggs sometimes tumble out from crevices and logs. In an effort to increase the available breeding habitat a nest box was designed. Looking across the gully there were several of these green plastic cubes installed among other nest sites. Short lengths of PVC pipe act as tunnel-like entrances to the boxes, wide enough to admit a Gould's Petrel but too small to give larger, predatory birds access.

Adult birds take it in turn to incubate the egg and a parent will remain on the nest for up to 21 days before being relieved by its partner. By the end of one of these stints the nesting bird can be weak with hunger and vulnerable to

attack. The nest boxes have proved to be a significant asset, both in terms of improving the petrel's breeding-success rate and allowing the researchers to study their charges at close quarters. The study program's Technical Officer, Nicholas Carlile, lifted the lid of one of the boxes and we peered in to see a petrel sitting snug and secure in its haven of polyethelene.

By day the gullies where the petrels nest appear calm, unruffled places. Were it not for the wire markers and the condominiums of green boxes there would be no visible sign of the nesting birds. Later that evening, however, the scene was transformed. Adults, returning from feeding far out to sea, circled over the island and then plummeted awkwardly through the forest canopy.

We too blundered our way through the gloom to the middle of the South Gully and sat while petrels clattered among the palm fronds. The night air was rent by the high-pitched cries of circling birds and the muffled growls from partners ensconced in their burrows. Adding to the babble were the hysterical calls of Wedge-tailed Shearwaters (*Puffinus pacificus*), who also have a breeding colony on the island during the summer months.

For such a delicate-looking creature and accomplished aerialist, the Gould's Petrel is ungainly on the ground. With wings outstretched for balance, they waddle among the rocks in search of their partners and nests. This rigmarole is then reversed in a few hours time when birds leaving the island to feed, stumble about, trying to select a gap in the canopy to fly through. Others scale the Cabbage Palms, using their clawed feet and sharp beaks to haul themselves up trunks and clamber through the fronds to a take-off point.

Adding to the petrel's woes are the Pisonia or Bird Lime Tree (*Pisonia umbellifera*), whose sticky fruits ripen at the time when fledglings are beginning to leave the nest. The fruit clings to the bird's wings and inhibits flight, leaving them at risk of starvation or attack. Initially the Pisonias were so dense that Dr David Priddel, the head of the conservation program, described the masses of fruitfall as "walls of death." A large number of the trees within the colony's boundaries were poisoned. But they

were not the only factor causing the population decline.

Intruders

The following morning the gullies were tranquil once more. I worked my way along the main ridge to the island's northern point. A warm draft scattered through the forest, sending an old Cabbage Palm frond crashing to the ground near my feet. A few paces further on I found a dead petrel, with its wings splayed across a boulder and head wrenched back at a grotesque angle. This individual had most likely been killed a few nights earlier when it collided with a tree during the chaos of rush-hour traffic.

But nearby I glimpsed two more elements in the mortal equation affecting the Gould's Petrel: a Pied Currawong (*Strepera graculina*) braying in the canopy and then, among the slanting shadows of the forest floor, three plump rabbits quietly grazing. As part of an ill-advised biological experiment, a single pair of rabbits was introduced to Cabbage Tree Island in 1906. Free from predators, the rabbits have prospered on the island, at the expense of the forest understorey. They have also munched away on seedlings of the larger trees that would have helped rejuvenate the tree canopy, including young Cabbage Palms.

With the steady thinning of their protective forest cover, the Gould's Petrel has come under increasingly attack by sharp-eyed intruders like the Currawong and Ravens (*Corvus* spp.), who frequent the island from their mainland strongholds. In a further move to bolster the survival rate among petrels, the Currawong population on Cabbage Tree has been culled with dramatic results[4]. But this is a stop-gap measure. Removing the rabbits poses a far greater challenge. Nevertheless, until this is done, the future of the Gould's Petrel and the integrity of the forest on which they depend remains problematic.

I pressed on through the forest and emerged at the top of the 100-metre-high cliffs that dominate the Cabbage Tree's east coast. A humid wind raced across the sea from the north-east, where Broughton Island lay low on the horizon. Dense cliff-top clumps of Mat Rush (*Lomandra longifolia*) were buffeted by the updraft. The more exposed ledges were colonised by gruesome-looking clumps of Prickly Pear (*Opuntia stricta*), another of the island's exotic pests and one as notoriously intractable as the rabbit.

Behind me I watched a Wompoo Pigeon (*Ptilinops magnificus*) feasting on fruits in the crown of a Sandpaper Fig (*Ficus fraseri*). A vision splendid in rich shades of green, yellow, deep purple and grey, this regal bird was once common as far south as the Illawarra region of New South Wales. Loss of habitat has greatly restricted its range to isolated forest pockets, such as those on Cabbage Tree, where the contests between the sacred and profane are sharply defined.

About Face

The following day we headed out to circumnavigate the island. The forested slopes soon disappeared behind the northern headland and our small inflatable boat was dwarfed by huge walls forming a ragged coastline of coves and rocky spires. An oily calm sea slopped against the island. Gulping sounds echoed from caverns and blow-holes as we edged along the shore.

Cabbage Tree, like neighbouring Yacaaba Headland, is a remnant outcrop of fine-grained granite and basalt formed by tectonic activity around 65 million years ago. In places along the shore, soft basalt dykes have been eroded away, isolating ribs of resistant granite. Elsewhere, pounding seas have helped cleave tall, jagged pinnacles from the main body of the island. The largest of these, Cathedral Rock, is surrounded by a moat of dark water and stands proud like an alpine aiguille. From an eyrie near the summit of this spire, a pair of Peregrine Falcons (*Falco peregrinus*) swooped and soared in fearsome displays.

Bobbing beneath this formation, there was no inkling of what lay on the opposite face of the island. These cliffs, and Boondelbah Island to the south, deflect the worst of the weather that assails Cabbage Tree, creating a lee slope where the forest is sheltered from rampaging

As a yacht slips by on a placid sea, the storm-battered ramparts of the island's east coast betray the Pacific Ocean's more frantic moods.
Quentin Chester

southerly gales. We rounded the southern point of the island, passing slabs and granite knolls where, the previous evening, we had watched groups of Little or Fairy Penguins (*Eudyptula minor*) come to shore, their bodies iridescent in the strobe-like flashes of the storm.[3] Like the Gould's Petrel and Wedge-tailed Shearwaters, they too are gully tenants, nesting in rocky burrows between forest and shore.

Petrel Head

On our final morning on the island I accompanied Nicholas on a routine tour of the North Gully. High in the gully we passed small Cabbage Palms Nicholas is nurturing - each seedling neatly fenced off from the voracious hopping herbivores.

Like a concerned midwife doing the rounds of a maternity wing, Nicholas inspected each

Cabbage Tree Island's southern ridge features rocky slopes interspersed with thick pockets of Mat Rush. Punishing seas have undermined the cliffs at the water's edge, creating caverns and blowholes.

nest. Reaching at full arm's stretch into hollows among the scree he carefully checked each burrow. Any petrel in residence was gently extricated for examination, as I transcribed details of band number, body weight and nest location into a notebook.

Recording such details is a repetitive task but the information gathered is critical to managing the survival of the Gould's Petrel. It was not hard to share in Nicholas's satisfaction when he discovered a fresh egg or a nest box newly occupied—small victories against the odds. For, in the end, this work is not just about protecting an endearing seabird, but restoring the vigour and natural sanctity of an entire ecosystem. The solution lies in our hands and in the palms of an island.

1   G.L. Werren and A.R. Clough, 'Effect of rabbit browsing on littoral rainforest, Cabbage Tree Island, New South Wales—with special reference to the status of the Gould's Petrel', in *The Rainforest Legacy, Australian National Forests Study*, vol. 2, G. Werren and P. Kershaw (eds), Australian Government Publishing Service, Canberra, 1991, pp. 257–277.

2   C. Davey, *A report on the numbers and distribution of Gould's Petrel* Pterodroma leucoptera *breeding on the John Gould Nature Reserve, NSW*, CSIRO Division of Wildlife and Ecology, Canberra, 1990.

3   Cabbage Tree and Broughton Island are the northern breeding limit for this species on the east coast of Australia.

4   In the 1994/5 breeding season the nesting success of eggs to fledge-staged chicks was around 70%, compared to previous outcomes of around 20%.

CABBAGE TREE ISLAND

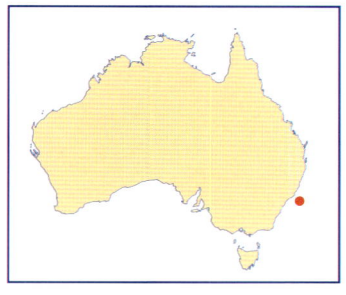

LOCATION

32° 42' S, 152° 14' E. North-east of the entrance to Port Stephens, 1.4 km from Yacaaba Head.

AREA

30 hectares

CLIMATE

Warm temperate/subtropical

STATUS

John Gould Nature Reserve, administered as part of the Myall Lakes National Park by the New South Wales National Parks and Wildlife Service.

ACCESS

No public access. Visitation is restricted to authorised scientific research.

FACILITIES

None

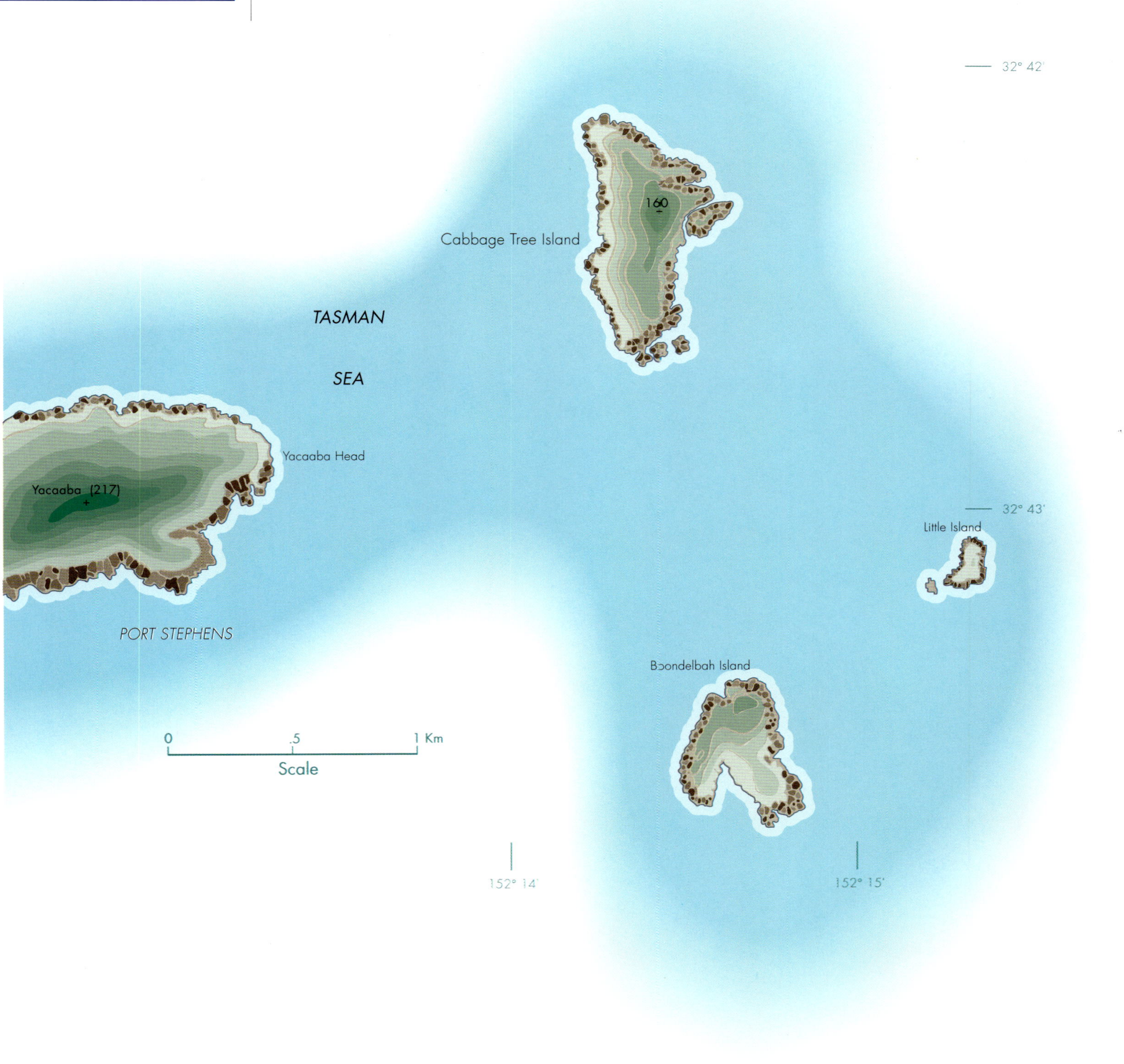

32° 42'

Cabbage Tree Island

160

TASMAN

SEA

Yacaaba Head

Little Island

32° 43'

Yacaaba (217)

Boondelbah Island

PORT STEPHENS

0   .5   1 Km

Scale

152° 14'

152° 15'

SPECTACLE ISLAND

Islands In The Stream

On the lower reaches of the Hawkesbury River there are around 20 islands. Many are tiny islets formed by drifts of alluvium swept downstream from nearby scarps and the labyrinths of Blue Mountains. Others are citadels of sandstone and scrub which became disconnected from the adjacent shores when the river valley was flooded by rising seas 10,000 years ago.

Over the past two centuries several of the larger islands have been modified. One is linked by a causeway to the mainland. Another is the site for a hospital. Some have been cleared, their appearance transfigured by buildings and roads. However, one has been left. It is little-known and hardly ever visited, yet every day Spectacle Island is seen by thousands of passing travellers speeding across the Hawkesbury.

For five years this cast-off chunk of sandstone was a personal landmark. During that time my family and I lived two kilometres downstream on Dangar Island. We were part of a small community of exiles. Dangar had sandy beaches, tall eucalypts and engaging locals. It was a kind of sanctuary, one step removed from the dissonance of the city. And yet, for all its charms, Dangar has been domesticated by houses, gardens and wharves. Spectacle Island, by comparison, retains the air of an archetypal landscape.

Commuting by dinghy to the mainland each day, I would invariably gaze upriver to Spectacle's hooded cliffs and shaggy ridges. This view was a reminder that we lived amid an inchoate landscape where natural forces held sway. In late winter, bitter westerly winds bore downriver. On summer afternoons colossal clouds would often build up behind Spectacle Island, spreading a pall of inky darkness across the valley.

Spectacle's claim to wildness nevertheless remains ambiguous. It stands less than 300 metres to the east of the six-lane Sydney to Newcastle Expressway and one kilometre west of the Hawkesbury River Railway Bridge, the main northern rail link. Aircraft approaching Sydney from the north often pass overhead. The island is bounded on three sides by oyster leases and overlooks a busy stretch of river plied by fishing boats and pleasure craft. At low tide the width of the navigable passage separating the island from the mainland shrinks to less than 10 metres. Yet the island's proximity to the river's hectic thoroughfares only heightens the unruly character of the place.

Over the years I made pilgrimages to Spectacle's shores and climbed up to shaded nooks. The press of crumbling overhangs and lavish undergrowth gave the impression of a site primeval. If I sat at the back of these dark clefts, the bridges were out of sight. On blustery afternoons, when the trees rattled and hummed, the din of the traffic was lost to the wind. And there were moments when I could try to envisage the spirit of the place that prevailed when the Guringai people inhabited the riverbanks and these islands in the stream.

Sandstone Gardens

The stubbornness of Spectacle's terrain has been its salvation. Like so much of the cordon of bushland that surrounds Sydney, it proved inhospitable to the ambitions of the settlers. This landscape owes its vigour to the Hawkesbury sandstone that was deposited as sediments when the entire region was an enormous estuary 200 million years ago. After the passage of more than 150 million years these consolidated sandstones and shales (in places forming a mantle 300 metres thick) were slowly uplifted. As the land rose, rivers gradually gouged deep courses into the soft rock, creating ravines and broad valleys crowned with cliffs.

Weathering of the rock surface yielded a thin cover of sandy soils. Despite being low in nutrients these soils support a phenomenal profusion of plant life of more than 2,000

Opposite: Burgeoning afternoon clouds gather over Spectacle Island, a familiar sight during late summer as storm cells migrate towards the coast.

Below: During spring the island's forest floor is dotted with colour, including the velvet-smooth Flannel Flower. Rob Jung

Fallen banksia leaves, seed cones and angophora bark form startling ground-level arrangements across Spectacle Island's many stone pavements.

Opposite: Despite the thin, impoverished soils that have developed on the Hawkesbury sandstone, an astonishing profusion of plants has adapted to life in this rocky habitat. Most conspicuous is the Smooth-barked Angophora, whose flange-like roots clasp their way into narrow fissures along the sandstone ledges.

crimson on the bell-like flowers of the Fuchsia Heath (*Eparcis longifolia*) and Christmas Bells (*Blandfordia nobilis*). Nectar-feeding birds like the White-eared Honeyeater (*Meliphaga leucotis*) would roam the undergrowth while King Parrots (*Alisterus scapularis*) cackled in the canopy trees. All this activity culminated with the flowering of the Christmas Bushes (*Ceratopetalum gummiferum*). Whole slopes were tinged pale red, as though lit by a perpetual sunset.

Floods, Fires And Famines

For all their Babylonian splendour, these hanging gardens are often starved by long, cloudless summers. The meagre soils remain dry and many of the smaller flowering plants wither in the desiccating heat. Leaf litter crackles underfoot and the rumble of trains on the rail bridge is drowned out by the thrumming racket of cicadas. In some years, whole trees succumbed to the drought. But the Angophoras, whose buttressed roots moulded onto seams along rocky ledges, somehow survived. As summer progressed they slowly shed their bark, leaving contorted limbs mottled orange and grey, while lustrous red gum seeped from their trunks like thick blood from a wound.

In such a parched state the bush was vulnerable, and during our time on the river, we often spent anxious summer days watching smoke billow from the ridges up and down the river. The air was heavy with the smell of burning eucalypts. When the dry northerlies raged, we watched dumbfounded as firestorms swept up slopes and trees exploded into vaulting sheets of orange flame. Spectacle was, miraculously, left untouched by the horrific fires that engulfed the Hawkesbury in January 1994. At such times these islands can appear as oases in a wasteland of cinders and ash.

The blight caused by fire and drought can, however, change very quickly. Autumn rains frequently trigger a sudden greening of the bush. In some years the rains persist for many weeks, often with torrential deluges. Fed by a vast catchment, the river becomes bloated with run-off, its normally placid waters churning in

species, making it one of the world's most diverse floral communities. The varied aspects and topography of Spectacle is a microcosm of these vegetation habitats. Steep rocky slopes encircle the island. This perimeter is dominated by the Smooth-barked Angophora (*Angophora costata*), Grass Trees (*Xanthorrhoea* spp.) and Coast Banksia (*Banksia integrifolia*), with a shrubby understorey.

On the drier northern ridges these species are intermingled with Scribbly Gums (*Eucalyptus haemastoma*), *Banksia serrata* and She-oaks (*Casuarina* spp.). The island's summit plateau is covered by an open woodland of Blackbutt (*Eucalyptus pilularis*) with heath species at ground level. Tidal mudflats on the north-western side of the island are colonised by mangrove communities, most notably the Grey Mangrove (*Avicennia marina*) and River Mangrove (*Aegiceras corniculatum*).

Visiting Spectacle on spring mornings I would meander up the gullies to the summit. As the island warmed in the early sun, the dank chill from the river disappeared and the bush came alive with the sumptuous colours and perfumes of wildflowers. Drifts of Flannel Flowers (*Actinotus helianthi*), like white star-bursts, dotted the understorey. Ribs of stone protruding from the vegetation were fringed by Rock Orchids (*Dendrobium speciosum*), with long bracts of creamy yellow flowers.

Elsewhere there were startling splashes of

silty, dun-coloured eddies. During lulls in the storms I occasionally took my boat upriver to Spectacle.

Under cloud-laden skies the island appeared drab. But in the dripping bush, everything seemed enriched: the sandstone faces in lustrous charcoal greys and the Angophoras glazed with hues of amber and apricot. Somehow, too, the brilliance of an Azure Kingfisher (*Ceyx azureus*) swooping along the water's edge was undimmed.

The biggest hazard was debris being bullied along by the surging torrent. Whole trees, wrenched from their foundations, became snagged among the rocks on the islands. Rafts of household detritus—tables, floorboards, plastic buckets and bottles—all bobbed past, flushed from settlements somewhere upstream. The river took on a musty smell and often there were dead animals and rotting fruit caught in these floating snarls of rubbish. In the midst of this chaos the islands were stanchions of sanity.

Shattered Peace In Broken Bay

Such events have been a force in this landscape since its inception. The brute power of the river was well known to the Aborigines in the region. Their familiarity with this terrain went beyond the sporadic inundations that ravage the Hawkesbury. It even transcended the notion of a once-in-a-hundred-year flood. Archaeological evidence suggests their intimacy with the region spanned thousands of years, back to the time when islands like Spectacle were still coupled to the mainland.[1]

Two Aboriginal tribal groups occupied the lower Hawkesbury and shoreline around Broken Bay at the time of early European contact. The inland area upriver was the domain of the Dharug, while the coastal region, including the tidal reaches of the Hawkesbury, were occupied by the Guringai people. Their marine precincts were replete with assorted fish, crustaceans and shellfish. These foods were supplemented by root tubers and wild game.

Although much of the evidence of Guringai life is based on observations by European explorers and settlers, the river-dwellers did leave a record of their existence. Throughout the region are broad sandstone pavements engraved with exuberant images, among them human figures, fish and other wildlife, shields and ancestral 'heroes', some depicted in outlines spanning more than 10 metres.[2] There are several hundred of these sites, many found in clearings on rocky tiers overlooking the waterways of Broken Bay.

Among the engravings are images recording the arrival of European ships to these shores. On 2 March 1788, just six weeks after the landing of the First Fleet in Australia, Governor Arthur Phillip set out from Port Jackson to investigate Broken Bay. Leading a party travelling in three small boats, Phillip went in quest of freshwater and farming land. They explored Brisbane Water, Cowan Creek and Pittwater before reaching Mullet (Dangar) Island on the afternoon of 7 March. But persistent rain and dwindling provisions forced a retreat.

By June 1789, however, the search was resumed in Broken Bay. Forays up Mullet and Mooney Mooney Creeks yielded little. Then on 12 June the party located an opening beyond Long Island. The strength and freshness of the flow proved beyond doubt that they had found the elusive river. After briefly returning to Port Jackson for supplies, a third expedition established the extent of the river and its tributaries.

The isolation of the area meant it was several years before the agricultural potential of the Upper Hawkesbury was realised. By 1800, however, more than 1,000 settlers were occupying alluvial flats along the river. Five years later this population had nearly doubled. The fertile soils supported grain crops, livestock herds and fruit and vegetables. Floods, however, posed a constant threat to the settlers. In eight of the 21 years between 1799 and 1821 the river rose by more than 13 metres.[3]

As settlement progressed, Aboriginal groups were driven from their traditional hunting and foraging grounds. In 1805 Governor King agreed to a Guringai request to stop the further spread of settlers. But the arrangement was short-lived. There were repeated conflicts and by the 1830s the Guringai presence was along

Opposite: The view from Spectacle Island's wooded eastern flanks takes in the waters of Cogra Bay and a ridge of Brisbane Water National Park on the mainland.

A band of weathered sandstone encircles Spectacle Island's flat summit, while the steep lower slopes are dotted with Smooth-barked Angophoras, Grass Trees and Christmas Bushes.

Island has withstood the turmoil of change. Over the past hundred years settlements have grown up on neighbouring shores and the river is now spanned by bridges. Following the Depression, many small shacks were built along the Hawkesbury. One such makeshift dwelling was perched on the south-eastern corner of Spectacle. Now only a few sheets of tin, some broken rubble and a midden of shattered beer bottles remain.

Inheritance

Not all my sorties to Spectacle Island were made alone. Occasionally my four-year-old daughter would travel with me, shrieking with glee as our dinghy leapt in the wake of cabin cruisers. She rode on my shoulders as we meandered among the grass trees and stared at the whiskery faces of the banksias. Kookaburras swooped past and startled wallabies would leap up and dissolve into the shadows.

On similar excursions along the bushy crown of Dangar Island we always stopped at a small bench of sandstone to look at the engravings of fish and wallabies. Brushing aside the scattered leaves, we would run our fingertips around the shapes. Being a four-year-old, my companion was full of questions. She wanted to know about the people who "drew" the pictures and "where they went." Part of my answer was to take her to another island, one unsullied by the confusing paraphernalia of streets and houses.

Being immersed in Spectacle's tangle of rock and scrub somehow made it easier to share what little I knew about the Guringai. We talked about their slender canoes and the way they speared their fish and hunted wallabies. We discussed how families sought shelter in the overhangs. But there were many quandaries of the past I felt powerless to explore. Instead I left my offsider to whirl wide-eyed through the bush. I let the island do the talking.

the river was limited to a few small outposts. The great bulk of the Aboriginal population had fallen victim to reprisals and disease.

Throughout the past two centuries Spectacle

1 A. Ross, 'Aboriginal Life on the Lower Hawkesbury at the Time of European Settlement', in J. Powell, L. Banks, *Hawkesbury River History*, Dharug and Lower Hawkesbury Historical Society, Sydney, 1990, p. 31.
2 G.L. Walsh, *Australia's Greatest Rock Art*, E.J. Brill/Robert Brown and Assoc., Bathurst, NSW, 1988, pp. 88-91.
3 R.I Jack, *Exploring the Hawkesbury*, Kangaroo Press, Sydney, 1986, p. 9.

SPECTACLE ISLAND

LOCATION
33° 32' S, 151°13' E. 40 km north of Sydney.

AREA
30 hectares

CLIMATE
Mild, temperate

STATUS
Nature Reserve within the Ku-ring-gai National Park, managed by the New South Wales Parks and Wildlife Service.

ACCESS
Private vessel via Brooklyn or Mooney Mooney. Public access restricted. Refer to the National Parks and Wildlife Service.

FACILITIES
None

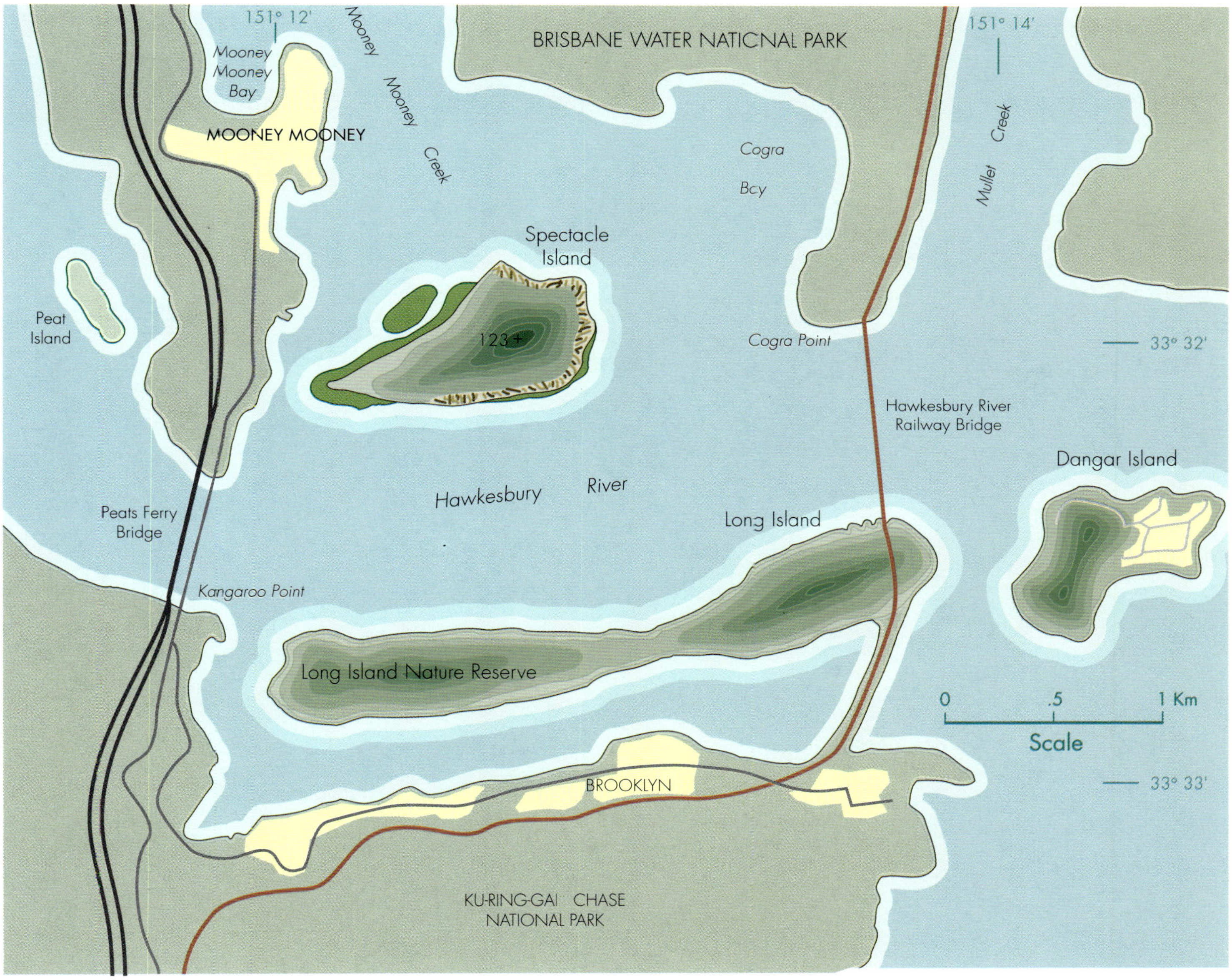

151° 12'

Mooney Mooney Bay

MOONEY MOONEY

Mooney Mooney Creek

BRISBANE WATER NATIONAL PARK

151° 14'

Mullet Creek

Cogra Bay

Spectacle Island

123+

Peat Island

Cogra Point

Hawkesbury River Railway Bridge

33° 32'

Peats Ferry Bridge

Hawkesbury River

Dangar Island

Kangaroo Point

Long Island

Long Island Nature Reserve

0   .5   1 Km

Scale

33° 33'

BROOKLYN

KU-RING-GAI CHASE NATIONAL PARK

Pacific Gull (Three Hummock Island)
Oil on canvas 107 x 91cm

BASTIONS OF BASS STRAIT

GLENNIE GROUP

KENT GROUP

HUNTER GROUP

Where long ago there stood mountain ridges yoking Tasmania to Australia, now only isolated summits remain. Encircled by unruly waters, these granite outposts have a turbulent history of shipwrecks, oppression and maverick plunder. In recent generations a natural order has been restored and a nobility regained. Yet, as the islands yield secrets of a deeper past, so they also endure as reminders of a harsh, reckless era.

GREAT GLENNIE ISLAND – GLENNIE GROUP

Beyond the Waves

The beach at Norman Bay on the western shore of Wilsons Promontory was crowded with families equipped with umbrellas, deck chairs and Eskies. At every turn there were tennis balls looping towards makeshift stumps and swimmers shrieking from the waves unfurling across the bay. Few in the crowd appeared to pay much heed to the surrounding granite peaks or the islands on the horizon. But occasionally a lone figure standing in the shallows would stare out to sea, as if transfixed by the enigma of what lay across the water.

We got into our kayak and, a few brisk paddle strokes later, we were clear of the bay. To starboard stood Norman Island, its massive rump of granite rising a hundred metres from the sea. Looking south, the rocky isles of the Anser Group gradually revealed themselves and on the far horizon hovered a dark peak, the natural fortress known as Rodondo Island. The only sounds were gentle ripples lapping at the bow and the occasional snort of a passing seal. The commotion at Norman Bay seemed a distant memory.

Granite Relics In Flooding Seas

Kayak travel gives one time to mull things over. In the case of Wilsons Promontory and its islands there is much to ponder. For a start, the rocks that are so evident in these parts have their origin in Devonian times. Around 390 million years ago, molten material pushed its way up beneath overlying deposits where it cooled to form granite. Subsequent erosion, spanning many millions of years, ground away the surface rocks to reveal resistant granite cores. Extensive weathering of these formations, in turn, eventually created the immense domes and residual tors that dominate the present-day terrain.[1]

No less thought-provoking was the fact that these islands are the exposed summits of an inundated mountain range. They are connected to that great submerged ridge that sweeps south-east across Bass Strait, through the Hogan, Kent and Furneaux island groups to Tasmania itself. When much of the world's water was trapped in polar icecaps during ice ages over the past two million years, this 50-kilometre-wide ridge was often exposed, bordering the Bassian Plain to the west. As sea levels rose after the most recent ice age 18,000 years ago, this expansive plain was flooded. Then 12,000 to 13,000 years ago the waters finally breached the land bridge connecting Tasmania with the rest of the continent.

Great Glennie returned to its island state not long after, around 10,500 years ago. While Aboriginal people continued to live on the larger Bass Strait islands for some time after their separation, the smaller islands lacked the resources to sustain a viable population. But small coastal outliers like Great Glennie that lay within striking distance of the mainland shores were visited on seasonal hunting trips.

In many respects this distant human connection was the most intriguing aspect to muse on as the islands drew closer. In choosing to travel by kayak we were in keeping with a tradition of canoe expeditions undertaken by the Brataualung people of Wilsons Promontory. By another coincidence we were also following in the wake of the first European vessels to visit these shores. They were not mighty sailing ships, but small craft powered by men pulling hard on oars.

Twilight Invasion

Gradually the rocky features of the shoreline became distinct. We steered into a quiet cove and in the calm of evening made camp on a tussock ledge just above the high-water mark. For the hour before dusk the island was a tranquil haven. The only evidence of animal life was Pacific Gulls (*Larus pacificus*). Yet even they

Above: Hoary Sunray, one of the common small shrubs of the island's tussock grasslands. Rob Jung

Opposite: Heralding the arrival of a cold front, freshening winds toss a lively sea onto the western shores of Great Glennie Island, with the domes of Dannevig Island and Citadel Island in the background. Rob Jung

stood motionless, like carved figures fixed to giant plinths of granite in the middle of the bay.

But then, as if in response to some celestial edict, swirling shapes suddenly filled the evening sky. Thousands of Short-tailed Shearwaters (*Puffinus tenuirostris*) circled overhead, forming an aurora of silhouetted bodies and outstretched wings. As the aerial invasion continued, a seaborne landing force announced its arrival with short barking calls that echoed across the bay.

Under the cover of darkness, parties of Little or Fairy Penguins (*Eudyptula minor*) emerged from the sea and gathered on rock slabs. Once above the slop of the waves they stood in small huddles like delegates at a convention. It was, as though, after a day devoted to underwater acrobatics, the transition to walking upright required an adjustment of attire and some conferring with colleagues.

Great Glennie is but one of many island strongholds for this species across southern Australia. Weighing about a kilogram, the Little Penguin is the smallest member of the penguin family. It is also the only penguin that is nocturnal on land. When darkness falls Little Penguins can be both aggressive and noisy; during the night we were treated to an array of penguin calls, ranging from throbbing growls to loud trumpeting. Together with the wailings of the shearwaters they created a surging chorus that made us nostalgic for the relative serenity of Norman Bay.

Northern Exposure, Southern Lights

Come daylight, however, the penguins and shearwaters had taken their leave of the island for another day in Bass Strait. We followed their example and launched the kayak, to paddle close to the shore.

The warm northerly breeze was rank with the island's heady seabird aroma. Approaching the northern point of the island a quartet of Cape Barren Geese (*Cereopsis novaehollandiae*) watched us from a steep, grassy slope criss-crossed with eroded paths. These plump grey birds look like story-book characters, misplaced on such a wind-swept outpost, but they remain an enduring symbol of Bass Strait.

Down the west coast of Great Glennie the northerly wind smoothed the sea, though the headlands and scoured expanses of rock were testimony to the gales that assail this shore. Our sedate boat speed was ideal for contemplating the immense monoliths. Along the shore, cliffs of smooth slabs disappeared into the water, plunging some 70 metres to the sea floor. In the course of the last ice age the base of these same cliffs formed the western rim of the continent.

After a leisurely hour under sail, we took up our paddles to push through the strait separating Great Glennie and Dannevig islands. Ahead stood Citadel Island, its granite flanks emerging from the sea like sheets of medieval armour. Silver Gulls (*Larus novaehollandiae*) squawked at us from their roosts among the coastal boulders. The remains of a haulage way led up to the summit and a small navigational light, which began operation in November 1913.

Despite the innocuous seas the only suitable landing we could find was a platform on the southern side of Dannevig Island. On its landward side the platform was surrounded by tall buttresses and slabs, draped with dense curtains of Rounded Noon-flower (*Disphyma australe*). The heat radiating off the granite amphitheatre was fierce. Penguins cowered in cool overhangs close to the water, while large limpets and liverish-coloured anemones shone in the warm rockpools.

From the minutiae of this inter-tidal universe I looked up to see an immense container ship on the southern horizon, passing between Rodondo Island and the Anser Group. Rodondo marks the boundary between east and west in Bass Strait. The lighthouses on Wilsons Promontory and Citadel Island watch over one of the busiest shipping channels on the Australian coast, a stretch of water the light-keepers call 'Bourke Street'.

"The Corner Stone"

The existence of a navigable sea passage between what were then known as Van Diemen's Land and New Holland was a matter of lively speculation during the early years of

British settlement in Australia. The first telling
evidence was provided by the surgeon and
redoubtable explorer George Bass. With six
crew he set out from Port Jackson on 3
December 1797 in a 28-foot (5-metre) whale-
boat.

Skirting the New South Wales south coast,
they passed landmarks familiar to Bass. Beyond
Cape Howe, they entered uncharted waters but
stiff gales delayed progress. Then, early on the
morning of 2 January 1798, Bass and his men
"were surprised by the sight of high hummocky
land right ahead, and at a considerable
distance." Of what was later to be named
Wilsons Promontory, Bass opined: "Its firmness
and vast durability make it worthy of being,
what there is great reason to believe it is, the
boundary point of a large strait, and the corner
stone of this great island, New Holland."

Strong winds and a breaking sea gave Bass
little time to savour this discovery. He steered
south, hoping to make Van Diemen's Land, but
the weather worsened and Bass decided to turn
back. The following morning the sea was still
too wild to reach the mainland but sighting
smoke on a nearby island (most likely Great
Glennie) they approached and were surprised
to discover it belonged not to "natives" but a
group of seven convicts.

These fugitives were part of group of 14 who
had escaped from Port Jackson in a stolen boat,
hoping to reach the wreck of the *Sydney Cove* on
Preservation Island. However, after having
failed to locate the wreck and its bounty, they
ended up on Great Glennie. Subsequently the
other members of the party had rowed off
under the cover of darkness, leaving the seven
to a dire fate marooned on the island.

Keen to continue his voyage, Bass gave the
castaways supplies and assured them he would
assist them on his return. After two weeks of
further exploration, which included the
discovery of Western Port and Phillip Island,
Bass and his men returned to Wilsons
Promontory. Gales hindered their progress, but
these delays allowed "petrels" and seals to be
gathered from the islands to add to their now
depleted rations. True to his word, Bass
returned to the seven convicts, dispatching five
of them to the mainland and taking two who
were ill aboard the whale-boat for the journey
back to Port Jackson.

Later that year another voyage by Bass and
Flinders would put the existence of the strait
beyond doubt. But, writing of Bass's earlier
foray, Flinders later noted: "A voyage expressly
undertaken for discovery in an open boat, and
in which six hundred miles [1300 kilometres]
of coast, mostly in a boisterous climate, was
explored, has not, perhaps, its equal in the
annals of maritime history."[2]

Stormfront

In contrast to the trials endured by Bass and his
crew, our morning circumnavigation of the
islands was decidedly tame. But soon after
returning to Great Glennie the wind swung
suddenly to the southwest. By sundown the

wind had built to a tremendous force and the seaward side of the island was pounded by a five-metre swell. Our previous night's camp was being battered by gusts fanning over the saddle.

For two hours we searched for an alternative site. In the end, we resorted to lashing our tents in the lee of a granite outcrop. As darkness wrapped around the island we sat watching the peaks of "the Prom" being engulfed by grey cloud. With each passing hour conditions appeared to worsen, but the returning shearwaters seemed unperturbed. Once more they blanketed the sky, deftly banking and sideslipping in the gale. Nor were the island's other regular commuters troubled by such events. Even though we were perched atop cliffs some 40 metres above sea level, it soon became clear that this was penthouse accommodation for Little Penguins. By torchlight we peered down the almost sheer drop to where adult birds were inching their way up steep ramps with astonishing skill.

Tussocks And Tors

The rocky ledges and crevices where the penguins nest are on cliffs that encircle the island. At higher elevations granite outcrops also provide valuable shelter to vegetation. The next morning, as we trekked along the island's southern ridge, we passed pockets of scrub with Coastal Tea-tree (*Leptospermum laevigatum*), Silver Banksia (*Banksia marginata*) and White Correa (*Correa alba*) huddled in the lee of large outcrops. There were moments when the bullying wind forced us to do the same.

Eventually even these monoliths are ground down by the elements. The grey, coarse-grained granite weathers to form the basis for the island's sandy, humus-rich soils. These soils are carpeted by Blue Tussock Grass (*Poa poiformis*) interspersed with smaller shrubs like Hoary Sunray (*Helipterum albicans*) and succulents such as Bower Spinach (*Tetragonia implexicoma*). Tussock grasslands cover most of the island's 66 hectares and their deep soil beds are honeycombed by shearwater nests. Behind huge cleft blocks on the ridge we found the abandoned nests of Cape Barren Geese, their downy feathers snared in thickets of vines.

During summer a shortage of grass forces the majority of these birds to leave the island and graze on mainland pastures.

As we sat atop a granite tor on the island's southern summit we were watched over by a White-breasted Sea Eagle (*Haliaeetus leucogaster*) soaring on the updraughts. These lords of the shore feed mainly on fish but their island smorgasbord would also include the small marsupial mouse, *Antechinus minimus maritimus*. Great Glennie is an important refuge for this endangered species, which, together with the Bush Rat (*Rattus fuscipes*), are the island's only two surviving indigenous mammals.[3]

From the summit we looked down on the narrow passage between Great Glennie and Dannevig islands. Huge waves toppled into the cliffs then collapsed in lathers of white foam. The placid waters we had glided through 24 hours earlier were now convulsed with the refracted swell. Through a gauze-like cloud of spray the unmistakable bastion of Rodondo Island towered in the distance.

To anyone with an eye for islands this is one of the most arresting sights in Australian waters. The island is girdled by sheer cliffs of red granite and its wooded upper slopes converge to a 351-metre-high conical peak. Rodondo was cut off from the mainland several hundred years before the land bridge to Tasmania was breached. This isolation and the island's distinctive topography have combined to preserve an ancient forest community unlike anything else in the region. It remained unexplored until January 1947, when John

A juvenile White-breasted Sea Eagle enjoying afternoon sun in the skeleton of a dead She-oak. Immature birds generally take five years to develop adult plumage and when fully grown their wing span can exceed two metres.
Quentin Chester

After a day's feeding, Little or Fairy Penguins gather along the granite shoreline before making an arduous ascent to feed their noisy offspring. By mid-summer the chicks are nearly fully grown and begin to shed their downy, grey-brown coats. At eight weeks the fledglings leave their nest and the adults depart for the open sea to fatten up for their annual moult.
Rob Jung

Bechérvaise fulfilled a boyhood ambition by leading an expedition that breached the island's formidable defences.

Between Two Worlds

After 48 tempestuous hours the gale began to abate. Emboldened by the thought of seeing Rodondo at closer quarters, we took to the water. Paddling in the lee of Great Glennie there was only the occasional windsquall. But in the open sea beyond Citadel Island we laboured to make headway against a three-metre swell. It was troubling enough to lose sight of land in the depths of the troughs. But the first thing we saw on the way back up was the spectre of Skull Rock, its immense cave looming like gaping jaws.

Given our faltering progress, and with no certainty of making a safe landing on any of the Anser islands, we turned back. With each roll of the swell, wave crests frothed over the bow and the kayak's wooden frame creaked around us. I felt a keen appreciation of the ordeal of Bass and his men, and that endured by the 14 convicts during their hapless quest along these

same shores.

Back on Great Glennie three hours later we sat enjoying the luxury of *terra firma*. In a small cave across the cove lay midden evidence that the Brataualung people made the most audacious journeys of all to the islands of Wilsons Promontory.[4] Over a period of 1500 years they crossed these treacherous waters, not in buoyant rafts, but in frail canoes made from sheets of eucalypt bark turned inside out and lashed fore and aft. As well as gathering shellfish, they would have been lured by the rich seasonal supply of shearwaters and seals.

The following day the winds eased and swung to the north-east. A raft of cloud blanketing the mainland was shunted south into Bass Strait. In brilliant sunlight we paddled back to Norman Bay and the world of gaudy beach umbrellas and ghetto-blasters. A few of the beachgoers welcomed us ashore. Standing in the shallows alongside our kayak, we talked of the raging winds and waves. I described our cliff-top camp and the sea and sky teeming with birds. But how to explain the visions of Rodondo and the island's intimations of ancient lives? At that moment, bestride two worlds, some enigmas seemed beyond words.

1   G. Wallis, 'Wilsons Promontory: An Introduction To Its Geology', *Victorian Naturalist*, vol. 97, 1980.
2   M. Flinders, *A Voyage To Terra Australis*, Facsimile edition, SA Govt Printer, Adelaide, 1989.
3   J.W. Wainer, 'Studies of an island population of *Antechinus minimus*', *Australian Zoology*, 19, 1976, pp. 1-9.
4   R.Jones, 'Bass Strait in Prehistory', *Bass Strait, Australia's Last Frontier*, ABC Enterprises, Sydney, 1987.

GREAT GLENNIE ISLAND – GLENNIE GROUP

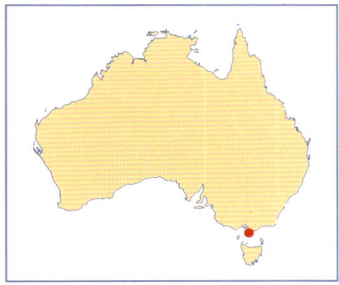

LOCATION

39° 05' S, 146°14' E. 6 km west of Wilsons Promontory.

AREA

66 hectares

CLIMATE

Temperate

STATUS

Part of the Wilsons Promontory National Park, administered by the Victorian National Parks Service. The waters immediately surrounding all islands south of Norman Bay form the Wilsons Promontory Marine Reserve. No amateur fishing is permitted and commercial fishing is to be phased out.

ACCESS

Restricted to beach areas only. Access for research or other purposes is by permit only. Contact the Park Information centre at Tidal River.

FACILITIES

None

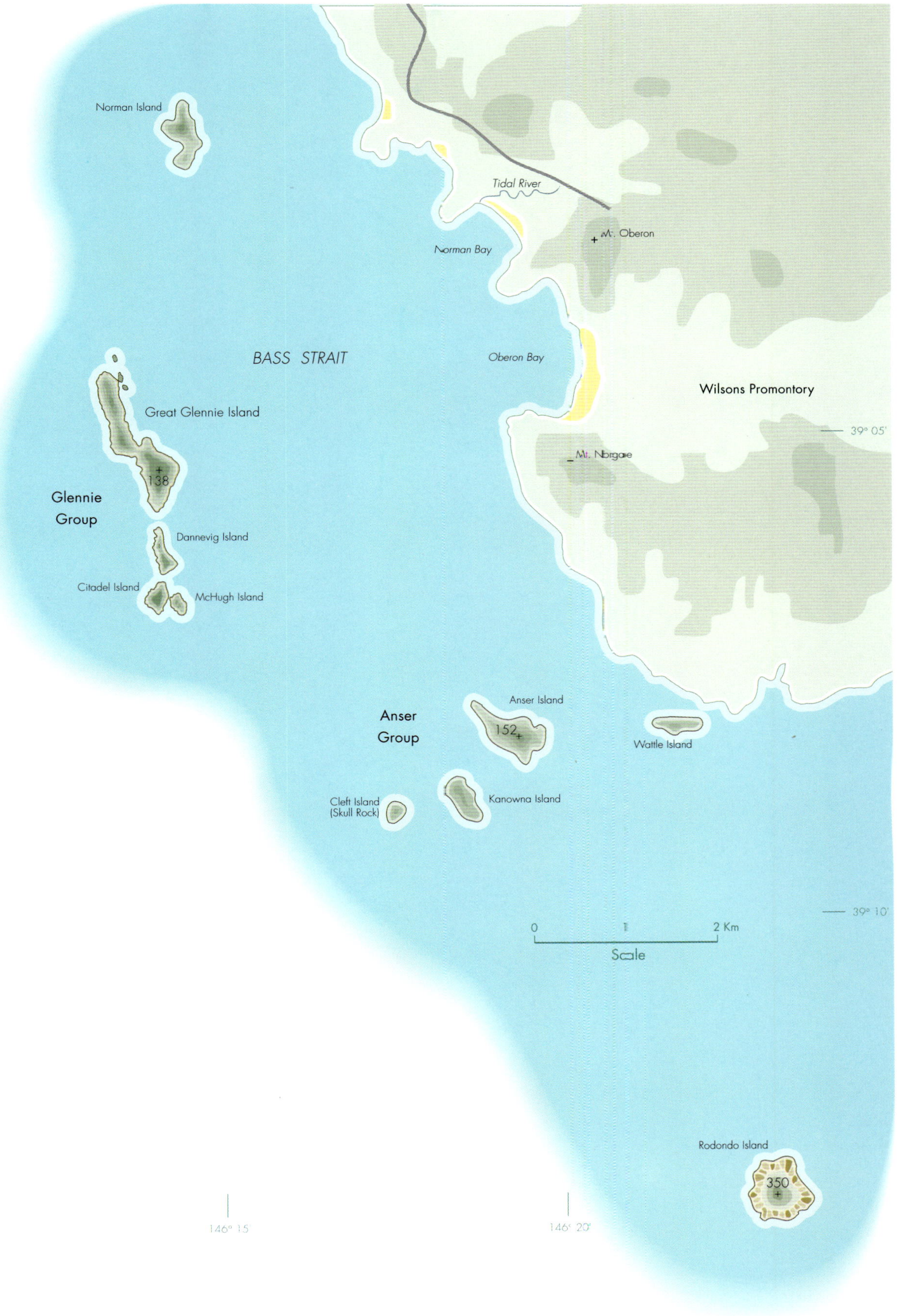

Norman Island
Tidal River
Mt. Oberon
Norman Bay
BASS STRAIT
Oberon Bay
Wilsons Promontory
Great Glennie Island
39° 05'
Mt. Norgate
138
Glennie Group
Dannevig Island
Citadel Island
McHugh Island
Anser Island
Anser Group
152
Wattle Island
Cleft Island (Skull Rock)
Kanowna Island
39° 10'
0
1
2 Km
Scale
Rodondo Island
350
146° 15'
146° 20'

Erith, suggests that "sand spits or other features may have existed, temporarily linking landmasses which are now separated by water."[4]

Sealing Their Fate

Whatever happened over the millennia, by the time Matthew Flinders arrived the Aboriginal inhabitants had long gone. As the nineteenth century unfolded, many more ships would visit the Kent Group. Some came in the name of science and discovery, others brought fortune

seekers, wild-eyed mavericks and men of God. These islands at the gateway to Bass Strait became a frontier. Here the forces that would change the face of an entire continent were acted out on a small stage.

Among those to follow Flinders was Lieutenant John Murray, who carried out a survey of the group in 1801 aboard the *Lady Nelson*. In these early years distinguished naturalists, like Robert Brown and Baron Ferdinand von Mueller, made important botanical collections during visits to the area. Brown's specimens of marine algae from the Kent Group were among the earliest and most

significant identified in Australia.

The Kent Group was a popular haunt during the heyday of sealing. George Augustus Robinson visited Deal Island in 1831 as part of his missionary quest to 'ameliorate' the conditions of Tasmania's remaining Aboriginal population, including many who had been brought to Tasmania's outlying islands. In Garden Cove he discovered "several rude huts or hovels, which are built of bags tied with hay bands, and thatched with grass." He also found gardens and boats belonging to the sealers, convicts and itinerants that used the Kent Group as a "rendezvous."[5]

Later Robinson helped establish a settlement at Wybalenna on Flinders Island, where in 1833 Tasmania's remaining 250 or so Aborigines were brought. The hapless experiment lasted 14 years. Those few Aborigines who did not die of European diseases were moved to Oyster Cove, south of Hobart.

New Deal

By 1847, when construction began on the Deal lighthouse, most of the sealers had moved on to new hunting grounds. Building a tower in such an elevated position was a formidable challenge. For months bullock teams hauled materials from the shore to the lofty site and convict labourers wrestled blocks of island granite into place. That the lighthouse, oil store and original keeper's cottage are still intact is a measure of their handiwork.

The cottage has now been put into service as a museum, where early photos of Deal show the island with dense vegetation and extensive stands of eucalypts. Over the years much of this timber was felled for use in construction, cleared for grazing or lost in bushfires. One blaze that swept the island in 1919 destroyed several buildings and the tramline to East Cove, prompting suggestions that the light be abandoned. Seventy-eight years later the changes proposed came into effect with the installation of the two outlying beacons.

The island has not, however, been left unattended. On a walk from Garden Cove to Deal's southern shores we called in on the new tenants. The lightkeeper's residence is now home to the Hollier family. They perform caretaking and meteorological duties but their real reason for being on Deal is not nautical but astronomical.

William Hollier is a physicist. His chosen field for research is self-replicating technologies. This branch of computing physics explores the theory that structures like robots and buildings can repair or replicate themselves. This is an idea that the United States' National Aeronautical and Space Administration (NASA) has a keen interest in. Deal Island was chosen as a site for this research because its isolation is seen as model for the remoteness of outer space, where it is hoped this new technology will be used.

In Deal's elemental setting of wind, rock and sea, the notion of space stations and mechanical devices with a burgeoning life of their own seems not just farfetched but totally alien. And yet, perhaps, the principle involved has its parallel in nature, where living organisms not only grow and replicate themselves but adapt to changes in their own habitat. Plants and animals do it—in a sense even islands do it.

Peppermints And Rednecks

Crossing the island, we passed the cleared airstrip and broad meadows of Blue Tussock Grass (*Poa poiformis*). One animal that has clearly benefited from the development of open grassed areas and has been busy self-replicating is the Red-necked or Bennett's Wallaby (*Macropus rufogriseus rufogriseus*). As we stepped among the tussocks and bracken, the wallabies leapt high to clear the vegetation and bounded away to the shelter of nearby trees.

The Kent Group is the northern limit of the Tasmanian subspecies of the Red-necked Wallaby. Most of the mammals endemic to Tasmania are also found on the Bass Strait islands. This distribution reflects the success of these animals in occupying the land bridge as it developed northwards during the last glacial period. By the time Tasmania was connected to the mainland at Wilsons Promontory, the Tasmanian species had moved well north. This pattern is also apparent in the distribution of several birds, reptiles and even plants.

From the cliff-top above Little Squally Cove we looked back inland. Under the late afternoon light the sombre grey of the casuarina scrub contrasted with the bright green of the Shining Peppermint Gums (*Eucalyptus nitida*), occupying sheltered ridges and valleys.

The Shining Peppermint is the most widespread of the eucalypts found on Bass Strait islands. Compared with examples of the same species on the mainland, it has smoother bark, larger buds and fruit, and different juvenile leaf shapes. As in the evolution of vertebrate fauna, these eucalypts became genetically isolated from their mainland relatives during earlier inter-glacial periods.

Standing proud above the tangle of vegetation was the lighthouse, stationed on the craggy promontory that separates Squally and Little Squally coves. Around our vantage point, those small shrubs that had kept their footing among the rocks were laid over and sheared into compact shapes - topiary by tempest. In the cove below, granite pinnacles, hewn by the elements, rose from the shore like the spires of Patagonia.

Judgement Day

We motored out of Murray Passage on a balmy morning with only a scrim of high cloud in the sky. Thirty kilometres to the west of the Kent Group a hump of grey granite rose from the sea. Observing this monumental islet 200 years ago, Matthew Flinders thought it resembled an elevated seat of justice. He named it and the neighbouring outcrops, Judgement Rock.

Exposed to the full force of ocean swells, these rocks are wave-swept and mostly inaccessible. Even on the calmest of days the heave of the swell makes jumping from a dingy onto barnacle-encrusted ledges a leap of faith. Above the range of the tides the granite is scrubbed smooth and bare of vegetation. It's an inhospitable place, as stark as an Aztec fortress. But for Australian Fur Seals it's heaven

The breeding colonies for this species are restricted to Bass Strait and Judgement Rock is one of only five such sites in Tasmanian waters. Even though the sealing boom was over by 1820, Australian Fur Seals did not receive official protection till 1891 in Victoria and as late as 1923 in Tasmania. It is estimated that some 8,000 pups are produced annually and the population is now close to 30,000 individuals.

As we stepped carefully through the herd the air resounded with a medley of barking yelps and guttural hollers. Those animals that weren't dozing on sunny ledges watched us intently and shuffled away as we approached. Small depressions in the rock were filled with

A ubiquitous presence on Deal Island's inland slopes, the Tasmanian sub-species of the Red-necked Wallaby is distinguished from its mainland cousins by a darker grey coat and a more subdued, brownish-red neck.

Overleaf: At Judgement Rock it's always rush hour and the weeks from early November to mid-December are especially hectic. During this period, female Australian Fur Seals give birth, before mating again with one of the territorial males a week or so later. Pups are generally weaned at 10 or 11 months and take four to five years to reach puberty. Widespread hunting during the early decades of the nineteenth century saw their numbers drastically reduced. Now a protected species, its population today numbers some 30,000.

The rocky fortress of South West Island is home to an automatic navigational light, one of the replacements for the now defunct Deal Island lighthouse. Encircled by granite cliffs, South West Island forms a high plateau with a tarn and a distinctive plant community unlike others in the Kent Group.

The rocky fortress of South West Island is home to an automatic navigational light, one of the replacements for the now defunct Deal Island lighthouse. Encircled by granite cliffs, South West Island forms a high plateau with a tarn and a distinctive plant community unlike others in the Kent Group.

urine and blood and occasionally we came upon dead pups that had been flattened by the crush of bodies during territorial struggles.

On the summit block—the judgement seat— I tip-toed past slumbering calves whose pups gazed up at me with dark, globe-like eyes. The waters below seethed with bodies; heads and tails breaking the surface in tumbling, exultant displays. Away from the shoreline young seals rose vertically out of the open water and tilted their heads back to bark a declaration. Others would sniff the air then rocket out of sight into the depths, leaving a fizzing trail of bubbles in their wake.

Watching all this, it was impossible not to reflect on what had happened barely two centuries before. Even at the time there was disquiet, not just at the annihilation of a resource, but the brutality of it all. Flinders bore witness to the "terror of the sailor's bludgeons" and the young cubs that "moaned piteously" in the face of the onslaught.

While there is good reason to take heart from the recovery of seal populations in Bass Strait, their future survival requires vigilance. Seals still die gruesome deaths. Among those we observed at Judgement Rock, a few wore necklaces of polypropylene net and the blue plastic bands discarded from fishermen's bait boxes—seals doomed to strangulation. Less obvious, but perhaps more significant, are the effects on seal numbers caused by the degradation of the marine environment and depletion of their food sources.

While Judgement Rock is a wildlife reserve, the status of the Kent Group and the management of Deal Island remains, at the time of writing, in limbo. The opportunity exists to create a significant marine and island conservation area. However, there is also a risk that the Kent Group could become a neglected outpost. Past mistakes have a habit of being replicated. For these islands stationed at the very crossroads of Bass Strait's natural and human history, the time has come to do more than just sit in judgement.

1   M. Flinders, *Observations on the coasts of Van Diemen's Land, on Bass Strait and its islands, and on part of the coasts of New South Wales*, John Nichols, London, 1801.

2   As quoted in M. Horder, *Mariners are Warned*, Melbourne University Press, Melbourne, 1989.

3   Murray-Smith's fellow travellers included Jack Mullett, M.A. Marginson, Jack Jones, Ian Turner, Clifton Pugh and Fred Williams. The ending of Joanna Murray-Smith's novel *Truce* (Penguin, Melbourne, 1994)

explores, through the agency of fiction, her relationship to her father and these islands.

4   R. Jones, 'Bass Strait in Prehistory', in S. Murray-Smith, *Bass Strait —Australia's Last Frontier*, ABC Enterprises, Sydney, 1987.

5   N. J. B. Plomley (ed), *Friendly Mission: the Tasmanian journals and papers of George Augustus Robinson*, Tasmanian Historical Research Association, Hobart, 1966.

DEAL ISLAND – KENT GROUP

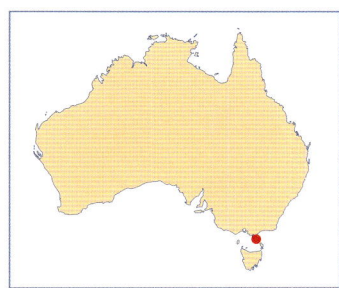

LOCATION

39°29' S, 147°21' E
Approximately half-way
between Wilsons Promontory
and Flinders Island at the
eastern entrance to Bass Strait.

AREA

2025 hectares

CLIMATE

Temperate

STATUS

Conservation Reserve
managed by the Tasmanian
Parks and Wildlife Service.
The lighthouse and associated
buildings are managed by the
Australian Maritime Safety
Authority (AMSA).

ACCESS

Via private vessel. East Cove on
Deal Island and West Cove on
Erith Island offer anchorages
for visiting yachts. Deal Island
has an airstrip for light
aircraft.

FACILITIES

No public facilities. A
caretaker currently occupies
the lightkeeper's houses in the
compound above East Cove on
Deal Island. One of the
original lightkeeper's dwellings
holds a collection of historical
material.

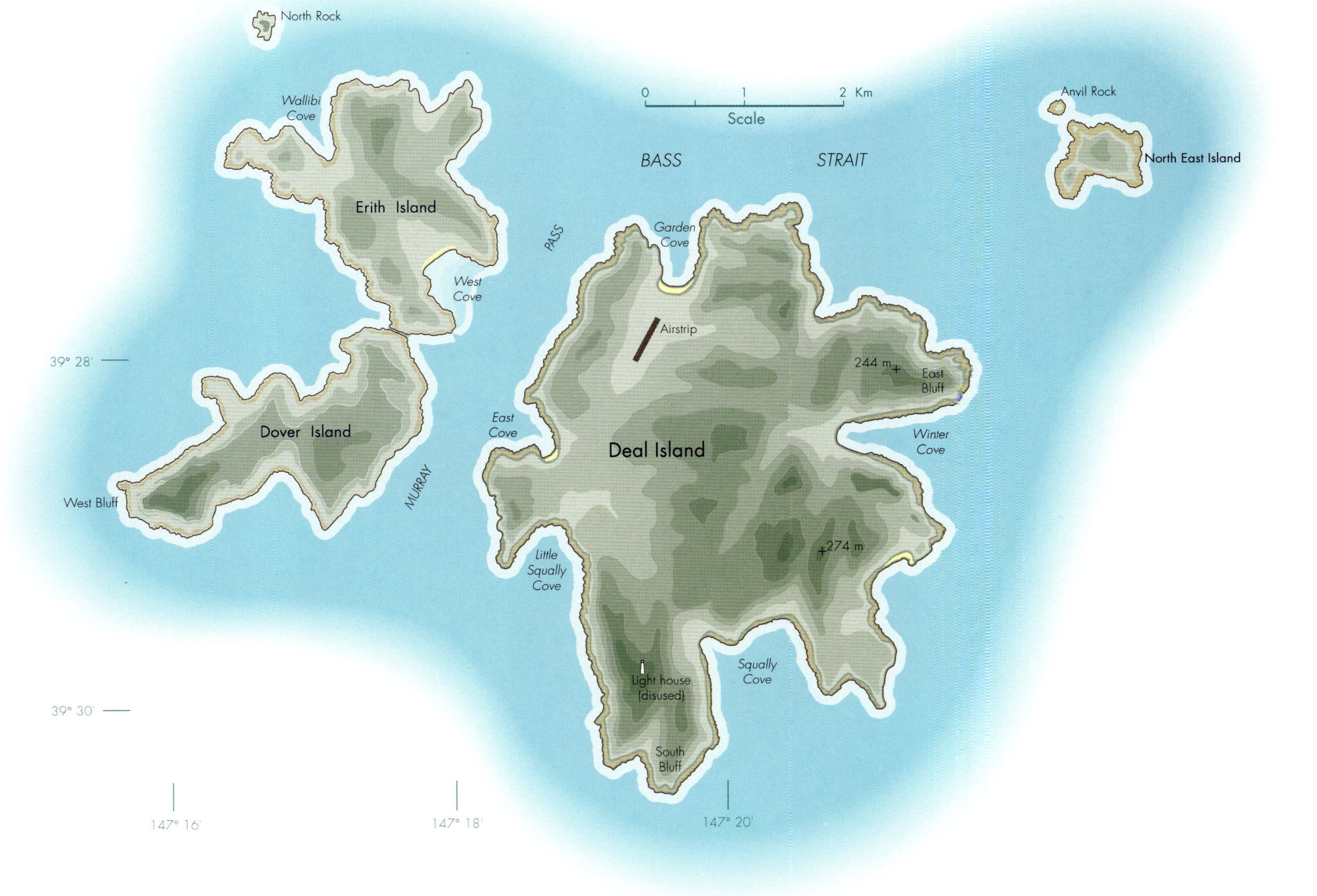

North Rock

Wallibi
Cove

Erith Island

West
Cove

Dover Island

West Bluff

East
Cove

Little
Squally
Cove

MURRAY

PASS

BASS   STRAIT

0   1   2 Km
Scale

Garden
Cove

Airstrip

Deal Island

244 m
East
Bluff

Winter
Cove

274 m

Squally
Cove

Light house
(disused)

South
Bluff

Anvil Rock

North East Island

39° 28'

39° 30'

147° 16'   147° 18'   147° 20'

Exiles

In 1951 John and Eleanor Alliston arrived at Three Hummock Island to begin a new life. They stepped ashore from the ketch *Jean Nichols* onto 9,300 hectares of tussock grass and wild bush. The local inhabitants included wallabies, tiger snakes, possums, 150 head of cattle, 400 breeding ewes and several thousand shearwaters. Broad sand beaches and granite bluffs encircled their new home. A boisterous 24-kilometre stretch of water stood between them and the outside world.

It was all a far cry from the life they left behind in England. John Alliston had spent 22 years in the Royal Navy. His commands included the destroyers HMS *Decoy* and HMAS *Warramunga*. At the end of the war the dislocations of naval life prompted a change of course. The Allistons yearned for continuity and a life they could share. "Our great purpose," wrote John Alliston, "was the journey: to travel together and raise a family under conditions of our own making."[1]

Over the years they toiled to realise their dream. They tended their sheep and cattle. Paddocks were fenced and farm machinery imported. An extensive vegetable garden was established. They found time to explore the lonely charms of their island and raise four children. Their isolation made all the vagaries of rural life more acute. Fickle prices, unsympathetic creditors, capricious weather— all had an impact. Yet they remained determined to preserve their cherished freedom. And 45 years later the Allistons continue to fulfil their "purpose" on Three Hummock.

A Late Intrusion

The islands of the Hunter Group are clustered together off Woolnorth Point, the far north-western tip of Tasmania. They are a disparate bunch. The largest island, Three Hummock, is roughly circular in shape. The terrain is hilly and timbered with Coastal Tea-tree (*Leptospermum laevigatum*) and large stands of Swamp Gum (*Eucalyptus ovata*) and Shining Peppermint (*Eucalyptus nitida*).

Just three kilometres away across Hope Channel lies slender Hunter Island, a flat spearhead of land. Much of this low-lying island has been cleared for cattle grazing. This, and its exposure to storms, has restricted the vegetation to coastal heath.

Robbins, the third of the larger islands, is similar in topography to Hunter. At low tide it is possible to wade to Robbins from mainland Tasmania. The main trio is accompanied by a dozen smaller islands and a scattering of islets and rocks.

The apparent differences between the islands stem, in part, from their geology. The rocks on Hunter date from more than 600 million years ago. They include quartzite, slate and siltstone which were folded to create a dome-like anticline. Prolonged erosion of this formation has left an elongated, gently undulating plateau around 50 metres high.

Three Hummock Island, by contrast, consists of granite intruded beneath overlying rocks 350 million years ago. Prolonged weathering of the granite bedrock has, nevertheless, produced an array of similar landforms, including scattered tors and dome-shaped outcrops. These processes also left three residual peaks—the conical summits that give the island its name.

The highest, South Hummock, stands at 237 metres. And it was in the shadow of this eminence that we landed from the boat *Wild Wind*. The site of our beach campsite was framed by rocky headlands. The coarse-grained granite is studded with chunks of quartz like peanut brittle. Erosion by wind and sea has left isolated blocks. At the tidal margins these blocks had been carved by wave action into fanciful, gargoyle-like shapes. In the deepening afternoon shadows the littoral turned figurative.

Bass Strait weather is rarely stable, and that

Above: Lashed by salt-laden winds, the predominant plant life on Albatross Island consists of lichens and succulents.

Opposite: An evening squall in Hope Channel buffets the granite bulwarks on the south coast of Three Hummock Island, near Neils Rock.

Evening light catches the granite foreshore of Rape Bay on the northern coast of Three Hummock Island.

night the inviting calm of our first day was obliterated by a riotous south-westerly change. The next morning we took refuge in the scrub, walking along tracks that led through dripping groves of Coastal Tea-tree and into the eucalypt woodland that cloaks the spurs of South Hummock. A history of grazing and fires may have left Three Hummock less than pristine but there are still expanses of bush where nature holds sway. From the saddle below Bronzewing Hill we looked north-east to thick scrub huddled around East Telegraph Bay. The sky was quilted with grey cloud, that brushed the island's three high points.

The Passage Betwixt

It was the trilogy of rounded hills that caught the roving eye of Matthew Flinders as he and

George Bass sailed along Tasmania's northern coast aboard the 25-ton colonial sloop *Norfolk* in 1798. On the evening of 8 December they dropped anchor in a "small sandy bight under the northern hummock." With their ships rations now at a low ebb, Flinders and Bass went ashore in search of food. They returned empty-handed but noted that it was low tide in the bay and deduced that the flood tide approached from the west.

This observation went to the heart of their quest for it offered Flinders further proof "not only of the real existence of a passage betwixt this land and New South Wales, but also that the entrance into the Southern Indian Ocean could not be far distant."[2] Early the following morning the *Norfolk* rounded Cape Kerauleren at the northern tip of Hunter Island. There Bass and Flinders were confronted by open water and

. . . a long swell was perceived to come from the south-west, such as we had not been accustomed to for some time. . . but, although it was likely to prove troublesome, and perhaps dangerous, Mr Bass and myself hailed it with joy and mutual congratulation, as announcing the completion of our long-wished-for discovery of a passage into the Southern Indian Ocean.[3]

During their time on Three Hummock, Flinders and Bass noted evidence of recent human occupation in the form of abandoned fire places. Later reports suggest that during the summer months groups of up to 50 Aborigines, including men, women and children, visited the larger Hunter Group islands. While catamarans were most likely used to cross long stretches of open water, contemporary accounts indicate people often swam between the smaller outliers.

For the Aboriginal people of Tasmania's north-west the offshore islands were a rich source of food. Little was known about these excursions until 1974 when a site on the eastern shores of Hunter Island produced evidence of Aboriginal activity spanning 25,000 years. This site gave prehistorians unrivalled insights into the origins of the first Tasmanians and the world they inhabited.

Visitors At Cave Bay

Archaeologist Sandra Bowdler spent several months excavating deposits in the floor of the cave for which Cave Bay on Hunter Island is named. This work yielded an array of stone and bone tools, pollen samples and animal bones. Analysis of the material collected helps develop a picture of the climatic and environmental

Left: A Shy Albatross feeding its fledgling. Shy Albatrosses mate for life, using the same nest each breeding season. A single egg is laid in September and after hatching the chicks spend increasing amounts of time alone while the adults roam the oceans to gather squid and jack mackerel for their offspring. After six months the juvenile birds make their first flight and do not return to the island for two years.

Opposite: The elevated terraces of Albatross Island are home to around 3,000 breeding pairs of Shy Albatross, estimated to be less than half the original population that in the early 19th century was plundered for its feathers.

upheavals that took place in those distant times.[4]

Around 23,000 years ago, when the cave was first occupied, the last great ice age was well underway. Sea levels were about 90 metres below their present height and falling, in response to the advance of polar ice caps. At this time Hunter Island was an outcrop on the exposed landmass known as the Bassian Plain, the bridge that linked Tasmania to mainland Australia. The nearest coast lay some 30 kilometres to the west.

In this phase the cave at Hunter Island was a base for hunting trips into the surrounding country. As the climate became progressively cooler and drier the moist forest around the cave was replaced with open, mostly treeless grasslands. The Cave Bay deposit reveals that this habitat was home to a variety of mammals, including species of bandicoot, kangaroo, wallaby, rodents and marsupial mice.

When the ice age reached its peak some 18,000 years ago, sea levels had dropped a further 40 metres. The arid plains to the north were raked by bitter winds. By this time, Aboriginal use of Cave Bay Cave had ceased altogether. The coast and its crucial supply of shellfish were some 75 kilometres distant. Water was scarce and vegetation sparse.

With the gradual warming of the global climate and the retreat of the glaciers that began around 12,000 years ago, sea levels rose rapidly and a vegetation cover of open woodland developed. Dating of midden layers in Cave Bay Cave suggests that occupation of the site did not resume until 6,600 years ago. It is thought likely that, even though by this date sea levels had reached their present heights, Hunter Island was still yoked to Tasmania by a ridge of sand dunes. This would have allowed the Aborigines access to the peninsula formed by Hunter Island.

The eventual removal of this sand bridge by wave and tidal action isolated Hunter Island. Another layer in the midden record suggests it was a further 2,000 to 3,000 years before the cave was revisited by sea-going foraging parties, establishing a pattern of seasonal occupation that lasted until the European invasion 200 years ago.

"White With Birds"

Sea-faring Aboriginal hunters would have had access to all of the Hunter Group islands except one: Albatross Island. It is extremely unlikely that their small craft would have managed to

The southern shores of Three Hummock Island are characterised by sandy bays enclosed by granite headlands and backed by steep tussock slopes riddled with the burrows of Short-tailed Shearwaters.

cross the 12 kilometres of rolling swell that separate this outlier from Hunter Island. When George Bass stepped ashore on Albatross Island on 9 December 1798, he was probably the first human in several thousand years to climb this rocky island, which Flinders described as being "almost white with birds."

On a blustery morning we set out from Three Hummock hoping to be among the few who have followed in Bass's footsteps. Rounding Cape Keraudren we were confronted by a three-metre sea and 25-knot south-westerly. Clouds raced low overhead, bringing misty rain. It was a regulation autumn day in Bass Strait. *Wild Wind* nosed into a small, protected cove on the north-eastern tip of the island. A lone Little or Fairy Penguin (*Eudyptula minor*) stood watching on the rocks as we rowed to the shore among kelp-clad boulders.

In a strait where the islands are characterised by smooth granite forms, Albatross is an oddity. The rock is a conglomerate, with rounded lumps of Precambrian quartzite fused together by dark basalt, like burnished creekbed stones set in tar. On the edge of the cove, the knobbly precipices around us appeared devoid of life. But on the island's flat-topped summit we came face to face with several hundred Shy Albatross (*Diomedea cauta*). With the indolence of youth, these fledglings sat preening themselves on

their nest mounds.

This species is the only albatross to breed so close to mainland Australia. The adult birds are among the least nomadic of the albatrosses. At the time of our visit the fledglings were almost fully grown and beginning to shed their downy plumage. While some 3,000 breeding pairs currently use this colony, it is estimated that the island was once home to nearly 20,000 birds. Those areas too steep or exposed for nests are colonised by thick rugs of succulent vegetation. In a sheltered cove on the island's eastern shores a bevy of seals was hauled out on rocks napping in the sun. On the opposite side of the island a steep sea chopped onto dark buttresses. Spray sailed over the island, buoyed aloft by a keen westerly.

By mid-afternoon the parents were riding this wind home. With a wing span of up to 2.75 metres, the birds were visible several kilometres away, gliding across the undulating swell. Soon, the sky over the island became layered with soaring shapes. Sharply tapered wings scythed the air. Bird after bird touched down around us. It was like sitting on an aircraft carrier, but the only sound was wind rushing over wings.

The Shy Albatross is superbly adapted for oceanic life. Yet they rely absolutely on their island home. Their rock is a place for mating and raising offspring. The island gives them the elevation they need to become airborne. Here the force that gives them their freedom always blows.

"A Pestilential Place"

When George Bass went ashore on Albatross Island he did so, not so much to admire the birds, but in search of food. He returned with his boat laden with seals and birds. Flinders observed that the albatrosses ". . . being unacquainted with the power and disposition of man, . . . did not fear him," adding, "we taught them their first lesson of experience."

It was a lesson that was soon to be repeated on a much larger scale. By the time the missionary George Augustus Robinson visited in 1832, sealing gangs were hunting not only their usual prey but the albatrosses as well. Their method was to drive the birds off their nests and send them tumbling down into the island's main gorge. This place was known as The State Prison, for once trapped in the confines of its sheer walls, the birds were unable to use their broad wings to get airborne. The sealers clubbed their captives to death, while in a cavern nearby the sealers' women, surrounded by piles of rotting carcasses, plucked the feathers. To Robinson it was "a pestilential place" where "the stench was intolerable." He reported that the sealers took as many feathers as would fetch a hundred sovereigns. According to one estimate, this would be nearly 4.3 tonnes of feathers. To make a single boat load would have meant the slaughter of approximately 3,200 albatrosses.[5]

Hunter, Hunters And Hunted

The depradations of the sealers were not confined to Albatross Island. Itinerant gangs roamed across the Hunter Group and north to King Island. As seal numbers dwindled they turned their attention to other animals, including Red-necked Wallabies (*Macropus rufogriseus*) and Tasmanian Pademelons (*Thylogale billardierii*).

By the 1830s, settlements had become established on the larger outliers. Land was cleared and burnt. A forest of Tasmanian Blue Gum (*Eucalyptus globus*), with specimens 90 metres high, once dominated the south-eastern corner of King Island. It was soon replaced by open grassland. Relic specimens of Tasmanian Beech (*Nothofagus cunninghamii*), from dense forests that once covered the island, probably suffered a similar fate. By the end of the nineteenth century the endemic populations of Wombat (*Vombatus ursinus*) and the small, stocky Emu (*Dromaius ater*) were extinct.[6]

Not only sealers raided the islands. HMS *Beagle* visited the Hunter Group in 1838 as part of its survey of Bass Strait. During their stay the ship's company feasted on boatloads of crayfish, ducks, fish and shearwater eggs. They also set fire to Three Hummock, and one of the ship's mates noted the following morning: ". . . everything is burned down—there is nothing to look at, no birds, or flowers, and the very stars are dimmed by the smoke. . ."[7]

By the time the exiled Italian patriot Giuseppe Garibaldi called into Three Hummock in 1852, the island's natural order had been partially restored. En route from Peru to Canton as captain of the *Donna Carmen*, Garibaldi landed at East Telegraph Bay. His only Australian landfall had a lasting impact:

How often has that lonely island in Bass's Strait deliciously excited my imagination, when, sick of this civilised society so well supplied with priests and police agents, I returned in thought to that pleasant Bay, where my first landing startled a fine covey of partridges, and where, amid lofty trees of a century's growth, murmured the clearest and most poetical of brooks where we quenched our thirst with delight . . .[8]

A "Pudding Of A Bird"

At our camp on Three Hummock the night-time sounds were not murmuring streams but waves crashing onto the sand and the cries of Short-tailed Shearwaters (*Puffinus tenuirostris*). Every year many thousands of this migratory species occupy burrows on steep sandy headlands like the one behind our camp. Each summer Tasmania is visited by 18 million of the 23-million-strong world population. The only shearwater to breed solely in Australia, *Puffinus tenuirostris* is this country's most prolific bird.

Since the 1790s the shearwaters have also

An adult Shy Albatross banks on approach to landing after a day's feeding in Bass Strait.
Quentin Chester

been known as Mutton-birds. From the outset they were an essential food for the sealers, as they had been for the Aboriginal parties visiting the Hunter Group. As seal numbers dwindled, mutton-birding became an important sideline. By the 1830s more than 750,000 birds were being taken each year. This seasonal harvest became a thriving industry that dominated life on the Bass Strait islands for more than a century and continues on a small scale to this day.

Aside from the vast numbers of birds that were available, the precise timing and synchrony of their breeding greatly simplified the harvest. The supply of Mutton-birds appeared inexhaustible. As well as providing a source of fresh meat, the rendered fat of the birds was used to grease skids in saw-mills and skips in coal mines. The feathers and down filled bedding and padded furniture. Their stomach oil was used for lamp oil. In the words of Stephen Murray-Smith, they were seen as a "cut-and-come-again pudding of a bird." [9]

Grazing For Geese

For decades John and Eleanor Alliston struggled to stave off the threat of losing their island. When their bank delivered an ultimatum in the late 1970s an agreement was reached whereby the Tasmanian government agreed to buy back Three Hummock. In return, the Allistons could continue to live "in peace." The island is now a nature reserve with its own flock of sheep. This is not so much a concession to the Allistons as a strategy for conserving one of Bass Strait's most famous and formerly most endangered inhabitants, the Cape Barren Goose (*Cereopsis novaehollandiae*).

While the Short-tailed Shearwater was a mainstay in the diet of early European visitors in Bass Strait, the most prized game were the geese. For Matthew Flinders "they formed our best repasts." However, indiscriminate slaughter and the invasion of traditional grazing pastures by introduced livestock saw a rapid decline in goose numbers. Hunting continued until the 1950s. A 1959 survey of the bird's main breeding grounds in the Furneaux Group estimated the population had dwindled to fewer than 1,000.

The Cape Barren Goose is a large bird and breeding pairs need an extensive grazing area. On the offshore islands the birds graze on tussock, as well as assorted herbs and succulents. But they are also partial to pasture grasses, and so grazing of sheep on Three Hummock benefits the geese by promoting new green growth. Since becoming a protected species, the Cape Barren Goose has made a healthy recovery. The Bass Strait population is now 12,000–13,000 strong.

At the end of our stay on Three Hummock we visited the geese congregated on the pastures and airstrip above Chimney Corner. We also called on the island's caretakers. Both the Allistons are now in their eighties and still wedded to their island solitude. As we talked of the island and their farming days these veteran islanders gave the impression of being as reliant on their habitat as their neighbours on Albatross Island. For the Allistons, Three Hummock keeps their freedom aloft. Yet it cannot be subdued. As John said, "The island's bigger than anyone who's tried to live here."

1   J. Alliston, *Destroyer Man*, Greenhouse, Melbourne, 1985, p. 185.

2   M. Flinders, *Observations on the coasts of Van Diemen's Land, on Bass Strait and its islands, and on part of the coasts of New South Wales*, John Nichols, London, 1801.

3   *Ibid.*

4   S. Bowdler, 'Hunter Hill, Hunter Island', *Terra Australis* 8, Department of Prehistory, Research School of Pacific Studies, ANU, Canberra, 1984.

5   D. Dorward, *Wild Australia*, Collins, Melbourne, 1977, p. 36.

6   J.H. Hope, 'Mammals of Bass Strait Islands', *Proceedings of the Royal Society of Victoria*, 85, 1973.

7   As quoted in M. Horder, *Mariners are Warned*, Melbourne University Press, Melbourne, 1989, p. 125.

8   G. Garibaldi, *Autobiography of Giuseppe Garibaldi*, Walter Smith and Innes, London, 1889, vol.2, pp. 65–6.

9   S. Murray-Smith, 'Islands of Bass Strait', in G. Dutton (ed), *The Book of Australian Islands*, Macmillan, Melbourne, 1986, p. 81.

THREE HUMMOCK ISLAND — HUNTER GROUP

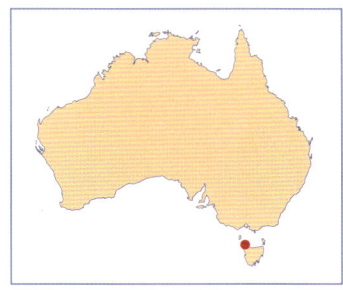

LOCATION

40°27' S, 144°55' E. 24 km off Woolnorth Point in north-western Tasmania.

AREA

7,110 hectares

CLIMATE

Temperate

STATUS

Conservation Reserve, administered by the Tasmanian Parks and Wildlife Service.

ACCESS

Private vessel only. Visits to the island by prior arrangement with the resident caretakers.

FACILITIES

None

Christmas Bells (Maatsuyker Island)
Oil on canvas 122 x 107cm

SOUTHERN OCEAN CASTAWAYS

MAATSUYKER GROUP

MACQUARIE ISLAND

NUYTS ARCHIPELAGO

ARCHIPELAGO OF THE RECHERCHE

Howling winds, huge seas and blasts of bitter cold—such are the moods of the world's most tempestuous ocean. Though the isles on the underside of the continent bear the brunt of these turmoils, they are anything but lifeless. For vast, shifting populations of seals, penguins and seabirds these are vital staging posts in their quest to survive and prosper.

MAATSUYKER ISLAND – MAATSUYKER GROUP

Round the Cape

We weighed anchor two hours before dawn and set sail for Maatsuyker Island, the site of Australia's southernmost lighthouse. It may be 352 years since Abel Janszoon Tasman first saw and named the island, but little has changed in this untamed quarter of Tasmania. The remote south-west remains as daunting as ever. From the sanctuary of Recherche Bay we headed into the darkness and mist. It felt as though we were sailing to the end of the line—to an isle at the brink of another world.

Maatsuyker is the outermost in a clutch of islands that lie roughly equidistant from Tasmania's South West and South East capes, two elongated ribs of rock that frame this coastline like parentheses. The island stands some 10 kilometres offshore, directly in the path of the notorious Roaring Forties. Above all else, Maatsuyker is defined by these fierce winds and the huge seas that whorl up from the icy expanses of the Southern Ocean.

At the same early hour as our departure from Recherche Bay, Owen Barret, head lighthouse-keeper on Maatsuyker, clambered out of bed and walked through the pre-dawn darkness to a small hut perched on the slope behind his house. There he collated some information—wind speed and direction, air temperature and barometric pressure—and phoned this through to the Hobart office of the Bureau of Meteorology, the first of four such weather reports from the island each day.

Meanwhile, the diesel generators in a shack down the hill near the lighthouse thumped on through the night, giving the island its electricity supply. They power the two-bar radiator in Owen's weather hut and cool the bottles of home brew in his refrigerator. The generators also drive a modest electric motor that keeps the lens of crystal prisms revolving slowly around a 1,000-watt lamp in the Maatsuyker lighthouse.

Since 1891 this navigational guide has been casting its cautionary beams to vessels passing to the south of the island and those making for Hobart from the west. Maatsuyker is one of only three lighthouses in a 240-kilometre stretch of Tasmania's southern and western coasts. The lighthouse and its custodians serve not only large passing ships, but are also a welcome beacon and reassuring human presence for fishing boats and small craft like ours plying the inshore waters.

We had been warned, at length, about conditions in these parts. Local fishermen and sailors spoke of the lack of secure anchorages, the fearsome gales and huge seas. There were stories of boats losing wheelhouses and breaking up. Other informants muttered about Maatsuyker's diabolical weather and the difficulty in landing. With all this in mind, and the forecast of a blustery change, our pre-dawn departure from Recherche Bay seemed only prudent.

But rounding South East Cape at dawn there was barely a whisper of breeze to fill the sails. The dark sea rolled by in glassy humps. To starboard, the coastline of the south-west straddled the horizon. Off in the distance the shoulders of Pindars Peak and Precipitous Bluff steepled into heavy cloud. Ahead a bank of low fog parted momentarily to give us glimpses of Ile du Golfe and De Witt Island, lit by the feeble morning light. Maatsuyker remained shrouded in milky grey mist, somewhere to the south-west.

The Cove

After an hour or so, I looked out to where the mist had parted and the vivid green flanks of Maatsuyker stood in a haze of light. By the time the yacht nosed into Haulage Cove, the sun had burnt off the fog and bright rays speared through the shallows to a sand patch where we dropped anchor. Bull Kelp (*Durvillaea potatorum*) writhed in the water and inquisitive seals swam around the dinghy while we ferried our packs onto sharp rocks.

Opposite: Fleeting showers race across the northern flanks of Maatsuyker Island, bound for Flat Witch Island on the left and the brooding bulk of De Witt Island in the distance.

Framed by wind-shorn Tea-trees, Walker Island is separated from Maatsuyker Island by a narrow rock-strewn opening. Louisa Bay on mainland Tasmania is just visible beyond Flat Witch Island on the right.

Haulage Cove is the only place resembling a protected bay on Maatsuyker. Following the decision in 1888 to build the lighthouse, the cove was used to land materials and supplies. A haulage way consisting of a two-way tramline was constructed up the steep slope from the landing. Trolleys laden with materials were raised by a horse-driven whim. The haulage way remained the sole means of access onto the island until helicopters took over the role of resupplying the island from the 1970s.

Over the years Haulage Cove has sported assorted jetties and storage huts. Each, in turn, has been swept away by storms. All that remains are the rusting struts of the lifting boom and the concrete platform where it stands, a favourite site for basking seals, who give the locality a rich, "blood and bone" smell.

For many centuries prior to its use by lighthouse-keepers the cove was a landing place for the Needwonne people who travelled from their coastal settlements around Louisa Bay on

shoreline, typically among large boulders and rocky hollows where pups and adult females shelter. Australian Fur Seals (*Arctocephalus pusillus*) are also a frequent presence. But the absence of suitable coarse gravel beaches and expanses of bare rock has deterred these seals from breeding here. Since 1970 a third species, the Southern Elephant Seal (*Mirounga leonina*), has been returning to Maatsuyker, re-establishing a small breeding colony in a secluded eastern cove—the only such colony in Australia, aside from those on the sub-Antarctic islands.[1]

Wild Westerlies

On the warm, windless morning of our arrival the aroma of seals was almost overpowering. It was a relief to climb the steep gully leading up to the grassy vehicle track skirting along the island's main north-western ridge. Tea-trees arching over the road created a verdant bower where Brown Scrub Wrens (*Sericornis humilis*) darted from bush to bush. At times it felt as though we were strolling down a country lane rather than traversing a gale-swept island.

Passing the helipad, we arrived at a garage workshop and were welcomed by David Clarke, Owen Barret's assistant, unloading tools from a Suzuki four-wheel-drive. We hitched a lift to the head light-keeper's residence where, for the first time, we spied the stocky white tower of the lighthouse and the rocks of the Needles.

Looking out of Owen's kitchen window, we could see the first signs of a north-westerly breeze riffling over the sea. The sharp profile of South West Cape reached across the horizon. It would be hard to imagine a kitchen sink with a more imposing outlook. Owen admits he never tires of the view. "It's always changing, something's always happening out there." For 11 of the past 12 years he has been Maatsuyker's lighthouse-keeper. His children spent their early childhood on the island, though his family now live in Hobart. As we talked, Owen thoughtfully stroked the ends of his clipped moustache. Like his lighthouse, Owen is well built and gives the impression of being a beacon of dependability.

The conversation turned, inevitably, to the

the mainland to gather shellfish and hunt for birds and seals. The crossings were made in small canoes constructed from bundles of tea-tree bark—bold undertakings given the prevailing winds and fearsome seas. A midden found at the cove in 1975 is thought to be around 570 years old.

In their quest for seals the Aboriginal hunters would have had a variety to choose from. New Zealand Fur Seals (*Arctocephalus forsteri*) breed at various sites along the weather. According to Owen, the placid conditions for our arrival were a bit out of the ordinary. "We probably only get three or four days a year like this," added David. The south-west is a famously cold, wet place. Maatsuyker has an average of 250 rain days each year. Despite this, Maatsuyker's average annual rainfall of 1,200 millimetres is less than half that of more elevated parts of the south-west. A moderating maritime influence also means that the island is subject to less extremes than the mainland.

But there is nothing moderate about the winds that buffet Maatsuyker. The island is exposed to a westerly airstream that flows uninterrupted across thousands of kilometres of open ocean. In winter, gales can blast Maatsuyker for a week or more. Fifty- and 60-knot winds are commonplace here. The highest recorded wind speed is 112 knots.

Muttonbird Mansions

By early evening we were ensconced in a campsite above a former helipad, one of the island's few patches of level ground. From the saddle below our camp we looked east to the small rocky isles of Flat Top and Round Top, while the steep south face of De Witt Island was lit by late afternoon sun.

Though no longer used by helicopters, the saddle's grassy clearing still serves as a landing strip. Just after dusk the sky began fill with the wheeling black shapes of Short-tailed Shearwaters (*Puffinus tenuirostris*), returning from their day's feeding at sea. For the next two hours thousands of these birds crashed to earth around us and then scrabbled through the undergrowth to find their burrows where hungry offspring awaited.

Maatsuyker is Tasmania's second-largest shearwater colony. Between September and April each year 500,000 breeding pairs occupy nesting burrows that are crowded into 100 of the island's 180 hectares. All this activity has radically altered soil structures and nutrient levels. Though the plant communities benefit from the dispersal of organic matter, the constant trampling by the birds causes significant erosion.

Into the night we were treated to cries and plaintive calls from conversing shearwaters. Judging from the volume of noise, several thousand seemed to be stationed by our tent. Then, around midnight, the clamour of the birds was overwhelmed by a south-westerly gale that tore up the ridge, buffeting our tent with rain squalls. By morning the shearwaters had taken flight across an angry sea. Waves cannoned into cliffs far below and the forest around us whistled and shook.

Tussock Grass clings to the steeply dipping bedrock of Maatsuyker Island's west coast, as seen from exposed slopes below the lighthouse.

Enchanted Forests

Maatsuyker Island is almost entirely covered by the Tea-tree (*Leptospermum scoparium*). On exposed ridges and cliff-tops they appear as little more than stunted shrubs, pruned low to the ground by the battering winds. But in protected sites they can grow to a height of six metres. Together with *Melaleuca squarrosa* and *Banksia marginata* they form a canopy so dense that, in places, less than six percent of the available light penetrates to ground level. While neighbouring islands, and much of the adjacent mainland, have suffered natural and deliberately lit fires, the flora of Maatsuyker has remained essentially undisturbed for nearly a century, thanks, in part, to the vigilance of the resident light-keepers.

To walk within the Tea-tree forest is to visit a self-contained world. I travelled north beyond our camp and into the trees crowning the ridge. The entwined upper limbs shuddered,

217

their branches clacking together like message sticks. Pairs of Green Rosellas (*Platycercus caledonicus*) gabbled as they foraged among the swaying canopy. In the shadows at ground level though, there was stillness. Ferns with glossy emerald fronds gave the forest floor a lush, primordial appearance. The ground rang hollow, every slope riddled with burrows.

As the ridge drops to the coast the Tea-tree gave way to low shrubs and sprawls of Blue Tussock Grass (*Poa poiformis*) and Pigface (*Carpobrotus rossii*). From the dark forest I emerged to blinding daylight and a vista of the ocean swell heaving onto Walker and Flat Witch islands. A cray-fishing boat was passing to the east, running with the sea. The rounded knoll at the island's northern extremity was ranked with dwarf specimens of Native Rosemary (*Westringia brevifolia*) and Correa (*Correa backhousiana*). Clumps of vivid yellow flowers clung to the fine gravel on the margins of the heath.

Along the western shore great fangs of rock rose from the sea. These sharply tilted formations, like so much of Maatsuyker's abrupt topography, reveal the steep dip of the island's 600-million-year-old metamorphic bedrock. The dominant rock types include

*Above: The vivid fruits of the Tasman Flax-lily (*Dianella tasmanica) *are a striking sight in the green on green bowers of Maatsuyker's upper reaches.*

Left: The more sheltered flanks of Maatsuyker Island are cloaked in a thick cover of Tea-tree which has survived unburnt for more than a century. The ferns and lush growth beneath the forest canopy provide a habitat for a large population of Swamp Antechinus. Meanwhile, below ground lie the nesting burrows of more than 500,000 breeding pairs of Short-tailed Shearwaters.

intensely folded and faulted Precambrian mica schists with exposed veins of quartz, and outcrops of phyllite and quartzite.

While waves bludgeoned the seaward faces of the rock-stacks, the inlet on their lee sides was sunny and strangely calm. Seals lolled on the slabs and frolicked in the sheltered pools. They snarled and bellowed with a repertoire of calls, at different times sounding like muted trumpets, barking hounds and shrieking horror movie victims—a surreal accompaniment to the wind thrumming through the undergrowth and the percussive booms of incoming waves.

Another boat passed by the outer rocks of the island heading south-west into the swell. Its familiar lines identified it as *Wild Wind,* a vessel often chartered by the Tasmanian Parks and Wildlife Service. Among those on board were Nigel Brothers and David Pemberton keen observers of the wildlife of the South-west islands for 20 years.

By late afternoon low-flying clouds brought intermittent rain to Maatsuyker. We spent the last hour before dark on the crest of the north ridge, gazing at these showers riding the wind to the ranges on the horizon. As they passed, the veils of droplets held the glancing light of

the late sun. Every now and then the distant peaks were framed in the sweep of a rainbow, stationary ribbons of colour in a sky rushing with cloud.

To The Lighthouse

The next morning was cold and grey. We sat huddled in rain jackets eating breakfast. From the cliffs below came the echoing sound of gunfire. In the space of half an hour there were a dozen more shots. The steepness of the island made it impossible to see what was happening. But the most plausible explanation was that fur seals along the shore were being used for target practice.

We strode back along the island to the lighthouse. Swamp Antechinus (*Antechinus minimus*) darted furtively in and out of the road's grassy verges. These small marsupial mice are the only terrestrial mammal to inhabit Maatsuyker and it is thought they may be a subspecies of this rare species. Feeding on insects and the island's only two reptiles, the Small scaled (*Leviolopisma pretiosa*) and Metallic (*Leviolopisma metallica*) skinks, the mice of Maatsuyker seem more relaxed about daytime activity than their mainland counterparts.

It is thought the absence of larger mammals on Maatsuyker may be a result of its relatively small size and lack of suitable habitat. Neighbouring De Witt Island, by contrast, supports Tasmanian Pademelons (*Thylogale billardierii*) and a large but fluctuating population of Eastern Swamp-rats (*Rattus lutreolus*).

While the land mammals of these islands live in relative isolation, their marine cousins roam freely among the outliers. From the cliff-top below the lighthouse we watched the Southern Ocean swell strike Needle Rocks. The largest of the waves sent great tusks of foam high into the air. Across the broad stone ramps of Seal Rock, the closest of these craggy islets, lay seals—perhaps a hundred—all seemingly oblivious to the uproar.

To the south-east were distant glimpses of the Mewstone, another rocky outpost besieged by tumultuous seas. This isolated rock-stack, like Pedra Branca further to the east, is also a popular haul-out site for seals and an important breeding colony for Shy Albatrosses (*Diomedea cauta*) and other seabirds. Pedra Branca is also the only known habitat of a small population of one of Australia's most endangered reptiles, the Pedra Branca Skink (*Pseudemoia palfreymani*).

Light Duties

After two hours in a bitter wind we took refuge in the lighthouse. A heavy timber door opened to a cosy haven of neatness. Stairs, freshly painted in gloss black and red, spiralled up to the light itself. The lens and its thick crystal prisms stand on a platform which completes a revolution every 180 seconds, casting its beaming message 26 nautical miles out to sea. The original clockwork mechanism is a marvel of cogs and counterbalances encased in a glass cabinet. Every surface in this working museum was spotless—the brasswork gleamed, the windows as clean as Windex could make them.

From the outset, the light-keepers on Maatsuyker have been much more than custodians of the light. In October 1907 six survivors of the wreck of the Swedish Barque *Alfhild* landed on the island. These men were fed and clothed by the light-keepers' families until the supply ship came.

Over the years the Maatsuyker light-keepers have been a vital link to the outside world. Their role encompasses everything from spotting fires in the South-west, to assisting with

Head Lighthouse-keeper Owen Barret checks his meteorological instruments. In the waters beyond the lighthouse lie Needle Rocks, a favourite seal haul-out site.

search and rescues. They keep in contact with fishermen and cruising boats in the area. When the fishing boat *Cape Sorell* ran into difficulties off the south coast in 1980, the keepers helped maintain contact with the search vessels. According to one of the rescued fishermen, "If the island was unmanned we would have gone."

The vigilance of the keepers also extends to those who might take pot-shots at seals. Walking up the path from the lighthouse we met up with Owen in his weather shack. He hadn't heard the gunfire but he didn't seemed surprised. "Bloody idiots," he said, turning to look out across the metallic grey sea, "sooner or later they'll get caught." Gathering evidence of these offences is notoriously difficult but the light-keepers have helped in successful prosecutions.

Though much of the weather watching at Maatsuyker is routine, the light-keepers can provide a service no automated weather station could match. "Occasionally the Met Bureau

rings us up and says, 'We've lost the front, have you seen it?'" said Owen.

Some of the most sudden and spectacular storms to hit the island are electrical. Lightning conductors are conspicuous on the lighthouse. Where hailstones have struck its copper dome there are dimples as big as those found on some scotch bottles. One violent storm that hit five years ago wiped out power and communications. "I went up to check on the radio tower and there was literally nothing there, only a hole in the ground," recalled Owen. "It just disintegrated."

Listening to Owen's tales, it seemed hard to believe that there are longstanding plans to 'de-man' this island. As one of the area's minders in Parks and Wildlife commented, "those blokes are one of our most endangered species."

Into The Night

Leaving Maatsuyker can take several days. In our case we had to wait 36 hours for the waves to subside before the dinghy ride across Haulage Cove became something less than suicidal. After four days scanning the horizon from lofty ridges it was strange to be looking at the world from *seal* level. As we rowed out to our yacht everything seemed larger than life: the island, the pitching swell and the bewhiskered seal faces peering up at us from the dark water.

We sailed away at dusk, passing to the south of Round Top Island, its sea cave rising from the water like the mouth of a leviathan. Rafts of Short-tailed Shearwaters rode the swell, enjoying a final feed before returning to their island retreats. Then the radio crackled with a weather forecast advising that another change was expected that night. In these waters the intervals between fronts can be disconcertingly brief. An hour later the island was lost in the darkness astern. But every 30 seconds there was a distant signal, a reassuring blink of light in the midst of a lonely sea.

1   N. Brothers and D. Pemberton, 'Status of Australian and New Zealand Fur Seals at Maatsuyker Island, Southwestern Tasmania,' *Australian Wildlife Research*, no. 17, 1990.

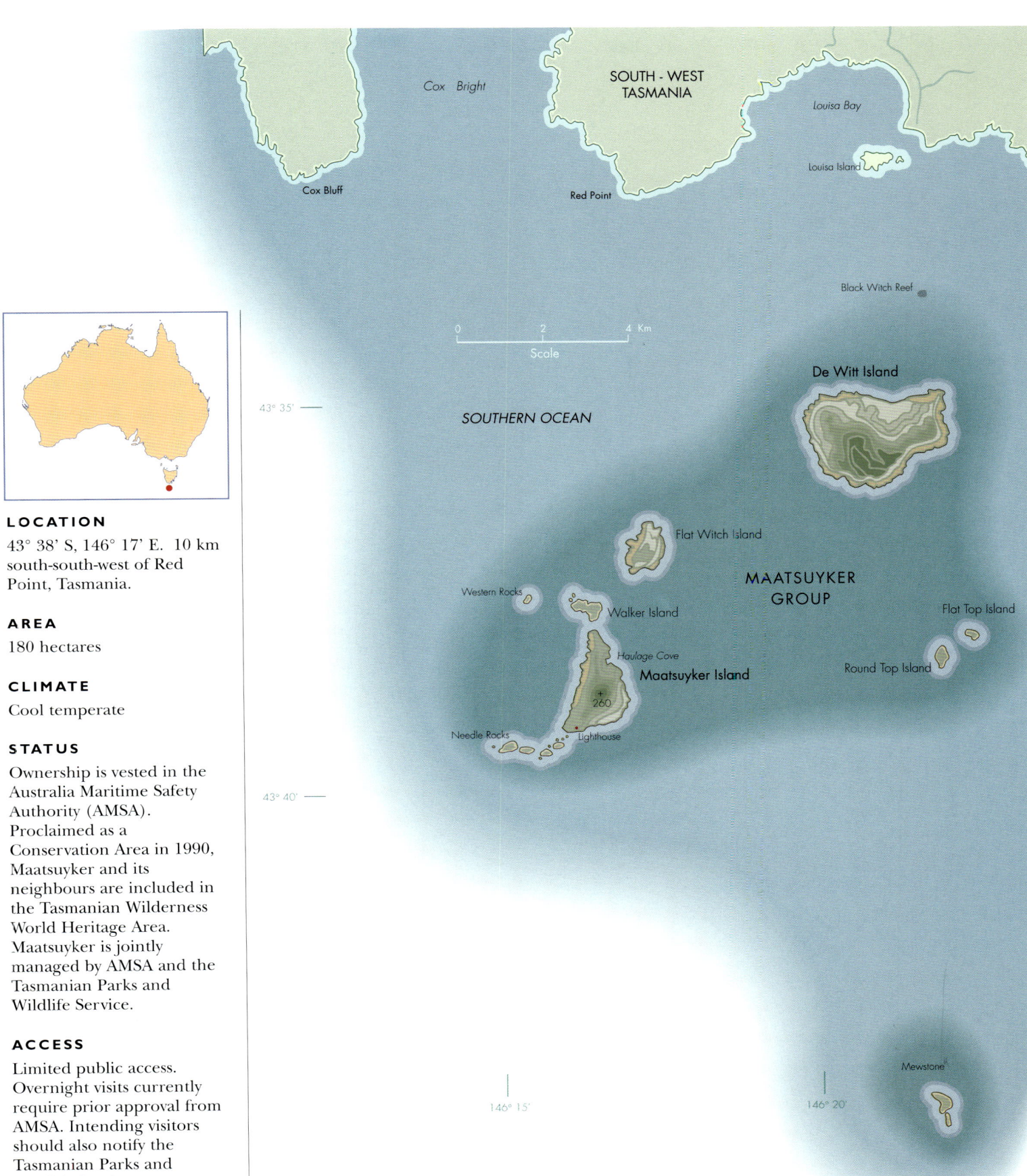

LOCATION

43° 38' S, 146° 17' E. 10 km south-south-west of Red Point, Tasmania.

AREA

180 hectares

CLIMATE

Cool temperate

STATUS

Ownership is vested in the Australia Maritime Safety Authority (AMSA). Proclaimed as a Conservation Area in 1990, Maatsuyker and its neighbours are included in the Tasmanian Wilderness World Heritage Area. Maatsuyker is jointly managed by AMSA and the Tasmanian Parks and Wildlife Service.

ACCESS

Limited public access. Overnight visits currently require prior approval from AMSA. Intending visitors should also notify the Tasmanian Parks and Wildlife Service.

FACILITIES

No public facilities.

MACQUARIE ISLAND

"... like a signpost pointing the way to the frozen lands"[1]

Three days into her voyage from Hobart, HMAS *Stalwart* easily rolled south. Drenched by fine spray, the lower aft-deck of the flagship of the Australian fleet became a cold and unpleasant place: I finally gave up. After many hours spent watching seabirds cavort over our churning wake, I turned for shelter within the cavernous warmth of the ship. *Stalwart* had crossed the most tempestuous waters on earth but the Southern Ocean had let us off lightly. It was almost anticlimactic when Macquarie Island hove into view early on our fourth morning at sea.

Light drizzle began to fall. A bleak Macquarie Island drifted in and out of the chill mist, as the ship came to anchor off Buckles Bay near the northern end of the island.

Our stay was to be brief. *Stalwart* had been hastily chartered by the Australian Antarctic Division to effect the resupply of the island and a change-over of personnel. At the time of our voyage, the Division's regular workhorse, the veteran Antarctic supply ship MV *Nella Dan*, was imprisoned by pack-ice far to the south, after attempting a marine science voyage early in the summer.[2]

The rain eventually stopped and the fog parted, revealing a smudge of watery emerald-green disappearing into clouds to the south. Most of the 34 kilometre-long island was still obscured from view and would remain that way for all but a few hours of my stay. Lying about half-way to Antarctica, 1,466 kilometres south-east of Hobart, Macquarie Island is Australia's most southerly sovereign territory. Like all the islands of the Southern Ocean, Macquarie's isolation is extreme. The closest landfalls are Campbell and Auckland Islands more than 600 kilometres to the north-east. A natural life-raft in the turbulent ocean, for nearly two-thirds of its brief history of human contact, the island was very much the opposite.

The dreary conditions that greeted our arrival were typical of the weather year-round. The island is almost perpetually wet, cold and windy.[3] Rain, sleet or snow can be expected on more than 300 days each year. A near-constant drizzle, rather than a torrent, is typical — fine soaking moisture that annually averages less than 1,000 millimetres. Snow can fall at any time but lingers only on higher ground and never accumulates in sufficient quantities to form permanent drifts. While flooding rains may not scour the Macquarie landscape, the island is exposed to the full assault of Southern Ocean storms. Gales, usually blowing from the west or north-west, assail its shores on more than 70 days each year. Maximum wind gusts can exceed 180 kilometres per hour.[4] Strong, nagging winds are part of the everyday Macquarie experience. Temperatures vary little throughout the year, with mean daily readings ranging between a chill 2.9° Celsius in August and a temperature less than 7.1° Celsius in March.

"like toothpaste from a tube"

Somewhere between 90,000 and 300,000 years ago, scraps of land emerged from the depths of the Southern Ocean atop an extensive mid-oceanic ridge. Extending from New Zealand towards Antarctica, Macquarie Ridge results from the massive forces of the Australian and Pacific tectonic plates pitted against each other. In the vicinity of Macquarie Island, the earth's crust has been squeezed together with such pent-up energy that the sea floor of the Pacific Plate has been warped, then thrust beneath the Australian Plate, resulting in massive uplift. As a consequence of the same process, just to the east of the island, the flanking Macquarie Trench has been driven down to depths of more than 4,500 metres. Intense earthquakes are common along Macquarie Ridge, indicating disturbances of the earth's crust on a grand scale.[5]

Of enormous geological interest, the island is

composed of a series of faulted volcanic and metamorphosed blocks — relatively pristine, exposed layers of ocean crust, showing much of the original sequence of seafloor formation. Described by geologists as an ophiolite, a rare fragment of oceanic crust above sea level, Macquarie Island has literally been squeezed up from the depths — as one researcher described it, "... like toothpaste from a tube".[6]

Estimates of age vary. Studies of the magnetism of rocks at the time of their formation suggest an age of 27–30 million years, although dating of crystallisation in basalts from several locations shows ages of 11.5 million years and younger.[7]

Indiscriminate Slaughter

As on so many islands in the Southern Ocean, Macquarie's human history began late, and unfolded with catastrophic consequences for the indigenous wildlife. It may have been visited by Polynesian mariners but no evidence of any landing or shipwreck has been found. The first recorded sighting was by Captain Frederick Hasselburgh on the brig *Perseverance*, while venturing well south from Sydney in July 1810. Financed by Robert Campbell, the prosperous Sydney merchant, Hasselburgh made several speculative voyages deep into the Southern Ocean. He discovered in turn Campbell and Macquarie Islands, naming one for his patron and the other for the governor of the colony of New South Wales, Lachlan Macquarie.

On discovering Macquarie Island, Hasselburgh landed a small party of men with a few months' provisions then hastened back to Sydney. But he failed to keep his discoveries secret and the island was soon being worked by numerous sealing gangs. Less than 20 years after the first European settlement in the region, this remote place became one of Australia's earliest industrial sites.

Macquarie, like all sub-Antarctic islands is home to a bounty of wildlife. A few tiny scraps of land in a seamless ocean, they teem with animals dependent on the sea. While the sea sustains life, the islands allow it to multiply — on a total of just 12,000 square kilometres of land in millions of square kilometres of ocean. Put simply, the islands act as scarce breeding and moulting platforms for marine mammals and seabirds that are otherwise totally unconcerned with terrestrial existence.

The shores of Macquarie Island are like a crowded zoo without the bars — animals everywhere. Before the arrival of humans, its rocky promontories and beaches would have been littered with more than 200,000 squabbling fur seals. Also piled on the beaches

Bull Elephant Seal. The largest of all seal species, the male Southern Elephant Seal can reach more than three tonnes in weight and grow to almost four metres in length. While males are sexually mature from about the age of six, dominant bulls will not attempt to dominate a "harem" of females before they reach 14. This still maturing individual already shows the scars of combat with other males.

were long steaming lines of Southern Elephant Seals (*Mirounga leonina*), hauled out to breed and moult.

Prized for their pelts, the fur seals were the first to be taken. So complete was their slaughter in the years immediately following 1810 that the original species distribution on the island remains unknown. The pelts were destined for the fashion markets of Europe. It was a lucrative but totally unsustainable trade. By 1821, barely 10 years after the island's discovery, fur seals were commercially if not zoologically extinct.

The Sub-antarctic and Antarctic Fur Seal (*Arctocephalus tropicalis* and *A. gazella*) did not resume breeding until the mid-1970s. Numbers of pups are still extremely low, with only about 100 born each year. A small colony of mostly "bachelor" New Zealand Fur Seals (*A. forsteri*) is also present, an overflow from colonies closer to New Zealand. Only one recent breeding record has been made. Whether New Zealand Fur Seals ever bred in numbers on Macquarie Island is not known.

Once the fur seals were gone, full attention was directed at the Elephant Seals and the extraction of seal oil. In 1820, Russian polar explorer Faddei Faddeevich Bellingshausen, the first scientific visitor to Macquarie Island, described the slaughter of Elephant Seals:

"One of the sealers accompanied us. He had with him an implement with which to kill sea elephants, which consisted of a club 4 1/2 feet [140 cm] long ... The end was bell shaped ... and bound with iron and studded with sharp nails. When we approached a sleeping sea elephant the trader hit him with the implement over the bridge of the nose ... [and it] ... lost all power of motion. Then the man took out his knife ... stuck it into its neck from four sides ... The animal then gave a few heavy breaths and died at once".[8]

A large proportion of the original population of perhaps 110,000 animals was exterminated from the more accessible parts of the island by 1830. The slaughter continued spasmodically throughout the nineteenth century but never with the same ferocity as in the early years. While the sealing gangs had all but eliminated the indigenous mammals, they

left in their stead a feral menagerie that included rabbits, cats, black rats, mice and the flightless Stewart Island or Maori Weka (*Gallirallus australis*).[9] Cats and wekas drove two endemic birds to extinction — the shore-dwelling Macquarie Island sub-species of the Red-fronted Parrot (*Cyanoramphus novae-zelandiae erythrotis*) and the Macquarie Island Rail (*Rallus philippensis macquariensis*), an endemic sub-species of the widespread Pacific Banded Landrail . Wekas, rats and cats greatly reduced numbers among several species of burrowing prions and petrels, while rabbits prospered. By the 1970s, there were thought to be more than 150,000 rabbits on the island, ravaging the vegetation.

"... the most wretched place of involuntary and slavish exilum ..."

Despite enthusiasm over the commercial potential of Macquarie Island on the part of ship-owners and merchants, life for those left on shore was hard. Captain Douglass of the *Mariner* called at the island in 1822, later describing it as "the most wretched place of involuntary and slavish exilum that can possibly be conceived".[10] Another colourful account of 1826 observed:

Seal or penguin oil awaiting shipment from Macquarie Island, c. 1900. Looking towards the Nuggets from Buckles Bay past the wreck of the Clyde.
Courtesy of Tasmanian Museum and Art Gallery, Hobart

Juvenile King Penguins, Sandy Bay. The birds in their down were born the previous summer and are still dependent on their parents. From courtship to fledgling may take 14 months or more. Only then can the juveniles go to sea and feed, no longer reliant on the adults. The long cycle of dependency means that one breeding season in three is missed.

"Parties belonging to two or three individuals are frequently living here at one time, and as keenly contested wars have occasionally raged between them for the dominion of a half mile of coast of this dreary purgatory ... and the combatants in their long beards, greasy seal-skin habiliments, and grim, fiendlike complexions, looked more like troops of demons from the infernal regions, than baptized Christian men ... the wretched stone and turf-walled and grass roofed hovels they inhabit ... send forth an odour to which that of the nightman's museum of foul abominations is myrrh and frankincense".[11]

Alongside the dwindling lines of seals, Macquarie Island's nineteenth-century coastline was crowded with stupendous numbers of penguins, particularly those massed in huge colonies of King Penguins (*Aptenodytes patagonicus*) and the endemic Royal Penguin (*Eudyptes schlegeli*). In a single colony, more than 70,000 pairs of Kings packed the shore at Lusitania Bay on the south-east coast.

The lingering oil industry was eventually monopolised by the Invercargill merchant,

of 160,000 pairs at Hurd Point.

"... one of the wonder spots of the world ..."

In the midst of this carnage the island was visited by some early Antarctic expeditions, including those led by Bellingshausen, Scott and Shackelton. A few scientific studies were made but the first extensive observations waited until the 1911 visit of Douglas Mawson. En route to Antarctica, he landed a party of five men who lived on the island until December 1913. While maintaining a vital radio link between the Australasian Antarctic Expedition (AAE) and the rest of the world, Mawson's men made observations of the island's geology, natural history and meteorology.

Mawson and the renowned photographer on the AAE, Frank Hurley, greatly moved by the beauty of the island and its wildlife, were also sickened by the killing. After World War I, both lobbied against Hatch's activities. Addressing the South Australian Branch of the Royal Geographic Society of Australasia in 1919, Mawson stated: "This little island is one of the wonder spots of the world ... it behoves those responsible for its good keeping to see that the animals resorting thereto are properly protected against any possibility of extermination."[12] Tasmania had controlled the island for nearly a century, so it fell to its government in 1920 to cancel all remaining oil licenses. The industry never recommenced.[13]

Mawson again visited the island in 1930 and continued to push for its protection. A permanent wildlife sanctuary was declared in 1933. Recent management efforts have sought to control rabbits, wekas and cats. The myxoma virus was introduced in 1978 and has been reasonably effective, reducing rabbits to a current population of about 10,000. Cats are shot and trapped but several hundred remain.

Since 1948 an Australian National Antarctic Research Expeditions (ANARE) Station has been continuously maintained at the northern end of Macquarie Island by the Australian Antarctic Division. Sited on a narrow neck of flat land known as The Isthmus, the station provides facilities for scientific research dealing

Joseph Hatch. In a hideous final chapter in the life of the enterprise, Hatch turned from Elephant Seals to penguins in an attempt to squeeze the last drops of profit from the island. In 1889, he had large cast-iron steam digesters erected at Lusitania Bay. At first, Hatch's men fed King Penguins to the digesters, rendering down tens of thousands of carcasses. The Lusitania colony was quickly decimated. Royal Penguins proved more "digestible", so Hatch spread his operations to other locations, including Nuggets Beach and the huge colony

Opposite: About 160,000 Royal Penguin pairs crowd the breeding colony at Hurd Point. With total breeding numbers somewhere between 810,000 and 960,000 pairs, Royals are the most numerous penguin on the island. The Macquarie population is one of the largest aggregations of seabirds anywhere.
Steve Trémont

with the island's natural environment. In addition, upper atmosphere physics is studied and meteorological data continually gathered, providing vital information in a region with few fixed weather stations. The population varies from 13 to 20 people during the winter months, increasing to about 40 over summer. Resupply ships usually call twice each summer to relieve personnel and restock the station.

The Crowded Zoo

Once ashore and ready for several days in the field, I headed south along the east coast from the ANARE Station in the company of two officers from the Tasmanian Department of Parks Wildlife and Heritage.

Their task was to continue work on an inventory of sealing sites and shipwrecks. The shores of Macquarie Island have claimed their share of wrecks. At least nine ships from the sealing era are known to have prematurely ended a voyage on the island or its adjacent reefs. Our ultimate destination was Aurora Point and the nearby Aurora Cave on the west coast.

On past The Isthmus, our progress was continually shouldered by great tussock-clad hills, many of them scarred by landslides. Sweeping south, ridge after parabolic ridge sprang from the narrow coastal margin, lively greens receding to pastel mauves and greys in the far distance.

At Half Way Hill, I climbed 50 metres to observe a small group of Light-mantled Sooty Albatrosses (*Phoebetria palpebrata*) nestled among streaming ribbons of grass. The most numerous of the four albatross species nesting on Macquarie Island, more than 2000 pairs of these superb, gentle birds, nest in small groups on steep slopes and grassy ledges. Accentuated with a quizzical white eye-ring and a turquoise line (sulcus) on the bill, a sooty-grey head grades with exquisite subtlety to an ashen breast and pale mantle. They are one of the most beautiful seabirds of the Southern Ocean.

A pair of prominent rocky pinnacles, known as The Nuggets, stood in constant view at the water's edge as we walked the first few kilometres south from Buckles Bay. At the back

of Nuggets Beach stood Hatch's digester tanks. Hissing valves, twisted tubes, levers and rivets — the machines of death were now but a silent rusty shambles. Ironically, Elephant Seals lounged among the debris, rubbing their sloughing hides against the digesters.

A small muddy stream crossed Nuggets Beach close to the digester. The stream's entire length was crowded with disorderly lines of Royal Penguins, sloshing their way to and from any one of six breeding colonies scattered from just behind the beach to a couple of hundred metres up in the hills. Those fresh from the beach were clean, their breasts a sparkling white after a krill-fishing trip, while the birds heading down were mud-spattered and filthy after running through the mire of the colonies. I fought my way through a chest-deep tangle of vegetation to a small rise within 20 metres of tens of thousands of penguins. Packed tight, each bird had just enough standing room. It was difficult to comprehend that this astounding mass of squabbling, odorous life was relatively small by Macquarie Island standards.

Being November, Royal Penguins were incubating eggs laid in late October. From the two eggs laid, only one chick would be raised, with the young birds fledging by the end of January. By May, the last Royals leave the island for the winter, not returning until September.

Back on the beach, we carefully negotiated the congregation of loafing penguins, trying not to trigger too much disturbance, then headed on towards Sandy Bay. The beach crest was strewn with rubbery straps of black and tan Giant "bull" Kelp (*Durvillaea antarctica*). Buzzing with tiny black flies, it had been pushed into thick rotting piles by the surf and now formed an ideal bed for somnolent

Elephant Seals. Rocks close inshore were collared by thick skirts of kelp, swirling with the push and pull of the swell. The head of a young Elephant Seal surfaced without a stir, snorted, then sank slowly back into the slithery tangle.

Approaching the first rocky promontory past Tussock Point, I noticed a commotion near the water's edge. An Elephant Seal lay motionless on the rocks, attended by a pack of Southern Giant Petrels (*Macronectes giganteus*) and Great Skuas (*Catharacta lonnbergi*). Moving closer, it became obvious that the seal was in fact a bloated carcass. The petrels, great lumbering birds the size of an albatross, with massive bills and stiff outstretched wings, squabbled and snapped as they jostled each other for space. Tearing at the hide, several disappeared up to their shoulders inside the carcass, each groping at the dead animal's viscera. Withdrawing, heads drenched in blood, they looked horribly like vultures of the sea. The island, such a bountiful place of life, is also inevitably one of uncompromising death.

The skuas impatiently stood their ground, pouncing on any scrap of flesh or offal spilt by the petrels, continually bickering over the spoils. A common predator as well as scavenger, the Great Skua's breeding cycle neatly coincides with those of penguins and burrowing petrels — undefended or abandoned chicks and eggs are among their main food items. Since the sealing era, rats and rabbit kittens have become important elements in the skua's diet.

At Sandy Bay, members of a small but growing colony of King Penguins were scattered along the beach. Juveniles in various stages of moult stood about, busily preening any accessible plumage. Some remained entirely covered in shaggy fur-like down, while others sported sub-adult feathers among motley brown tufts. Like Royals, King Penguins have recovered well from the predation of last century. More than 400,000 Kings now inhabit the island. The huge gathering at Lusitania Bay is the rival of those on South Georgia, site of the world's largest colonies.

Emerald Isle

By late afternoon any trace of watery sunlight had been enveloped by freezing mist pouring down the gully behind Sandy Bay. It was time to retire for the night to the welcome refuge of one of the six ANARE field huts scattered around the coast.

The following morning began as miserably as the previous had ended. Drizzle sent us on our way to Bauer Bay on the west coast. Cutting a deep path through the vegetation, the track up to the plateau was also a running stream. Its sides were clothed in dense and dripping foliage typical of the more protected parts of the coastal slopes, terraces and better drained valleys. Our feet were sodden within minutes.

Poa foliosa tussocks and clumps of the fern *Polystichum vestitum* were interspersed with patches of Macquarie Island Cabbage (*Stilbocarpus polaris*) and delicate mats of feathery-leafed *Acaena magellanica*. Said to taste like celery, the Macquarie Island Cabbage, a fleshy plant with geranium-like leaves and massed, globular yellow flower heads, was grudgingly consumed by sealers. As well as adding some balance to the monotony of seal and penguin meat, birds eggs and meagre ships rations, the vitamin C-rich cabbage kept scurvy at bay.

The island's flora is typical of the region: its complexity is closely linked to the severity of the environment. But in comparison to the more Antarctic Heard Island, the Macquarie flora is quite diverse and often strikingly lush.

Many species are shared with both its near and distant neighbours, and some are even shared with Australia.[14] Only three species are endemic — a low proportion, in part a result of the island's youth.

Altitude, drainage and degree of exposure to the elements are the natural determinants in the distribution of the five plant associations occurring on the island, but the extent and composition of some communities have been radically altered by rabbit-grazing.[15] Tussock grassland occurs on the better drained stretches of the coastal terrace, steep lower slopes and sheltered inland valleys. It affords protection from predators and is a major nesting habitat for burrowing petrels. On raised beach terraces and areas of high watertable or wind exposure below 400 metres, tussock grassland gives way to a shorter herbfield.

Towards the northern end of the island, large areas of the central plateau are dominated by herbfield, featuring, in various combinations, Macquarie Island Cabbage, *Acaena* spp., velvety brilliant-green cushions of the endemic *Azorella macquariensis* and spectacular rosettes of the widespread herb *Pleurophyllum hookeri*.[16] In the wettest, most poorly drained locations, fens and bogs prevail.

Our track climbed into the mist and across the exposed spine of the island. A break in the clouds afforded glimpses of a wild landscape of hills and solemn brown-grey peaks, smeared

A lone Wandering Albatross waits for its mate to arrive from sea. Pairing for life, Wanderers are at least nine years old when they commence breeding, and then only raise a single chick every other year. On the wing Wanderers have no equal. They soar and glide with ease in near continuous flight over thousands of kilometres of ocean, venturing as far north as the Tropic of Capricorn.

Above: The spectacular herbs Macquarie Island Cabbage and Pleurophyllum hookeri, *two species to have responded rapidly to the reduction in the island's rabbit population in recent years.*

*Opposite: Rockhopper Penguins (*Eudyptes chrysocome*). Rockhoppers are habitués of rocky foreshores and promontories, but their breeding colonies may be found well inland and up to 60 metres above sea level. The fertility of Rockhoppers has declined over recent years, with the Macquarie population of 3,000 pairs now half that of the 1970s. Other islands have also seen drastic declines. Campbell Island's Rockhoppers have dwindled by 94 per cent, or more than 700,000 pairs, since the 1940s. Rising sea temperatures and a resultant drop in food supplies have been implicated in this disturbing trend.*

thinly with green. On the flanking hillsides, gale and frost-sculpted natural terraces, neat steps of sparse vegetation and bare gravel, gave the vague appearance of a place more archaeological than natural.

The most conspicuous plants on the plateau's feldmark community are the compact, almost aerodynamic cushions of *Azorella macquariensis* and several of the 80 species of moss. In places, the moss *Rhacomitrium lanuginosum* forms irregular but distinct rows parallel to the prevailing wind. Large areas lie bare of all vegetation in the face of persistent gales. Towards its southern end, the plateau rises to more than 300 metres and is topped by the three highest peaks on the island, each over 400 metres.

We passed close to Square Lake, one of numerous freshwater tarns and ponds that stretch the length of the plateau, then descended to Bauer Bay and another welcome field hut. The west coast was its usual rowdy self. A strong westerly tugged at the manes of grass streaming from isolated pedestals of *Poa foliosa* tussock near the foreshore below the hut. Surf crashed around the numerous rocks and reefs trailing out to sea in chaotic lines, disappearing somewhere out among the mist and spume. Next morning we departed grudgingly in steady rain and strong wind for Aurora Point, about six kilometres to the south. Past Mawson Point, the narrow coastal plain was dotted with building sized rock-stacks, all draped in dripping vegetation.

The route had looked straightforward — just a matter of weaving around rock-stacks and crossing the open coastal terrace. However, we had not considered what are commonly known as featherbeds. Scientifically called by the somewhat more alarming name, quaking mire, these extensive areas of raised bog, topped by floating rafts of vegetation, are a feature of the north-western coastal terrace. Dominated by mosses, they often float on layers of peat several metres thick. Perhaps more like a waterbed, the unpredictable stretches of featherbed made progress slow and tiring.

With the morning gone, there was only time for a cursory look at the cave and its contents. A cramped cleft in a large rock-stack, its floor was littered with bird bones, many of them quite large and obviously those of the Wandering Albatross (*Diomedea exulans*). Sealers or ship-wreck victims had used the cave extensively. Further investigation would obviously reveal much of interest about its transient occupants and their prey.

Back at Bauer Bay in the early evening, the island became unusually calm. A couple of hours of glorious unexpected sunshine left time to wander around the foreshores, looking at the landscape and its inhabitants, in literally a new light. In the distance, the folds of the great coastal slopes rose like rumpled drapery, its creases modelled by the oblique sun. Small colonies of Gentoo Penguins (*Pygoscelis papua*) were scattered among the tussocks, chicks gathered in loose creches under the watchful protection of several adults. On bald-topped trampled hummocks, Gentoos stood calmly, each outlined in a lingering penumbra, the vaporous air suffused with light.

Our last day began as bleakly as all others. We decided to return to the ANARE Station via the coast and the broad sweep of Handspike Point on the north-west corner of the island — once again in depressingly familiar conditions.

During our stay, I had been hopeful of sighting that most magnificent of all seabirds, the Wandering Albatross. A small number breed on Macquarie Island. Almost at Handspike Point, I was resigned to not encountering any Wanderers, when just ahead a large white form came into view. It sat totally exposed to the blast, yet patient and serene, a single drop of water dangling on the point of a huge pink bill.

Nearly exterminated during the sealing era, Wanderers slowly recovered to a peak of about 50 breeding pairs by the mid-1960s. Numbers have been in decline since. Research in the 1980s established that Wanderers and other albatross species are now often hooked as an unintended by-catch in longline fishing operations over large parts of their range. It has been estimated that over 40,000 birds across a range of species are killed this way each year in the southern hemisphere.[17] Whether modifications to fishing techniques will stop the decline is not yet certain. Macquarie Island's tiny population of Wandering Albatrosses may again face extinction.

Metal and Plastic, Sinew and Feathers

More featherbed, and a biting wind around Handspike Corner before we were again within sight of Hasselborough Bay and the ANARE Station. Rather than sinew and feathers, a bird of metal and plastic rose over The Isthmus, then flew at a tilt into the westerly wind. The *Stalwart*'s helicopter was completing the last few resupply runs to the field huts. Just back at the station, I found myself bundled aboard the last flight down the coast. Strained forward, I stared out the open cargo door, instantaneously reviewing the terrain we had so laboriously walked. Recalling the geologists' claims, I was struck by something of a revelation. So

completely alone in a watery universe, it truly is one very special slice of ocean floor.

Our task complete, the mechanical bird roared back to Buckles Bay and then finally out to *Stalwart* for the voyage home. Passing over the Nuggets, the real avian life far below spotted the beach like confetti. Many thousands of penguins crowded around the few remains of Hatch's endeavours. In a heartening show of resilience, the beach and the island had long ago been reclaimed by nature. Macquarie Island continues to be that precious tethered life-raft, born of a rare fragment from the deep. With continued attention to its wounds and sustainable management of surrounding seas, the island must surely remain so.

1   J. S. Cumpston, 'Macquarie Island', *ANARE Scientific Reports* A(1), 93, 1968, p. 1.

2   Both ships are no more. HMAS *Stalwart* was decommissioned and scrapped in 1990. In December 1987 the *Nella Dan* became yet another victim of Macquarie Island's treacherous coastline. During unloading operations the ship ran aground at Buckles Bay in a gale. It was later decided to scuttle her, under controversial circumstances.

3   Compared to the severe and decidedly Antarctic conditions prevailing on the more northerly Heard Island (53°05'), Macquarie Island's (54°30') climate appears anomalous. The reason for a climatic inconsistency is not one of proximity to the South Pole but relative position in relation to the irregular circle traced out by the Antarctic Convergence. Macquarie Island lies slightly north, Heard Island more than 100 kilometres south. See page 261 for further explanation.

4   Mean windspeed for the year is 33.5 kilometres per hour. Meteorological data is gathered at the observatory at The Isthmus. Wind velocities on the plateau are at least 25 percent greater.

5   In 1980, nine seismic events registering more than magnitude five on the Richter scale were recorded in the vicinity of Macquarie Island. P. M. Selkirk *et. al.*, 'The island's origin and geology' in *Subantarctic Macquarie Island, Environment and Biology*, Cambridge University Press, Cambridge, 1990, pp. 42–58.

6   R. A. Kerr, 'Ophiolites: windows on which ocean crust?', *Science*, 219, 1983; quoted in Selkirk, *op. cit.*, p. 58.

7   See R. A. Duncan and R. Varne, 'The Age and Distribution of the Igneous Rocks of Macquarie Island', *Papers and Proceedings of the Royal Society of Tasmania*, vol. 122 (1), 1988. Also, P. E. Williamson, 'Origin, Structural and Tectonic History of the Macquarie Island Region', *Ibid*, vol. 122 (1), 1988.

8   F. Debenham (ed.), *The Voyages of Captain Bellingshausen to the Antarctic Seas, 1819–1821*, Hakluyt Society, Series II, 1945; quoted in Cumpston, *op. cit.*, p. 42.

9   Rabbits and Wekas were both deliberately introduced to Macquarie Island by sealers to supplement the meagre rations brought ashore and the monotony of seabirds, eggs and seal meat.

10   *Sydney Gazette*, 12 December 1822; quoted in Cumpston, *op. cit.*, p. 60.

11   P. Cunningham, *Two Years in New South Wales*, 1827; *Ibid.*, p. 67.

12   *Ibid.*, p. 311.

13   The industry ended as much from a decline in demand for seal and penguin oil with the increased availability of petroleum-based products, as from restrictions upon its operation.

14   Campbell Island, one of Macquarie's closest island neighbours supports a vascular flora of 128 species yet it lies only 2°12' north.

15   Palatable species such as the Macquarie Island Cabbage and *Poa foliosa* are thought to be significantly reduced over their natural range while some unaffected species have profited. Myxoma virus was introduced to the Macquarie Island rabbit population in the late 1970s. By 1993 numbers were estimated to have dropped to about 10,000. The subsequent recovery of the island's flora has been rapid and sustained. G. Copson and J. Whinam, *Rabbits and Vegetation, Their Future on Macquarie Island*, Tasmanian Parks and Wildlife Service, 1994.

16   *Azorella macquariensis* was identified as separate from the circumpolar species *A. selago* in 1989. See Selkirk, *op. cit.*

17   W. K. de la Mare and K. R. Kerry, 'Population dynamics of the wandering albatross (*Diomedea exulans*) on Macquarie Island and the effects of mortality from longline fishing', *Polar Biology*, 14(4).

MACQUARIE ISLAND

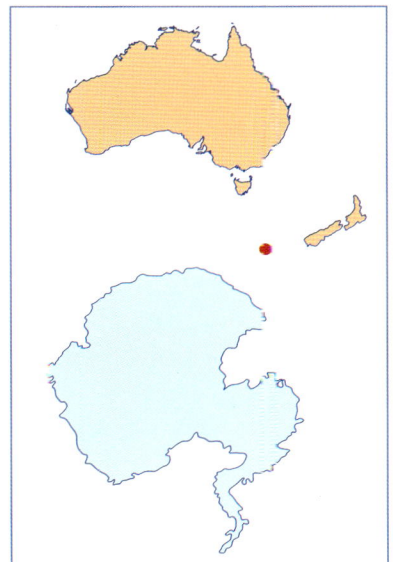

LOCATION

54°30' S, 158°57' E. 1,500 km
south-south-east of Tasmania and
1,466 km from Hobart.

AREA

12,800 hectares

CLIMATE

Severe—cold, wet and windy

STATUS

Tasmanian State Nature Reserve,
UNESCO Biosphere Reserve,
administered by the Tasmanian
Parks and Wildlife Service.
Logistic support and facilities by
the Australian Antarctic Division.
Nominated for World Heritage
List, 1996.

ACCESS

Regular resupply and personnel
change-over visits by ships
servicing the Australian National
Antarctic Research Expeditions
(ANARE) Station at the Isthmus.
Intermittent visits by tourist ships
and private vessels. Permit
required to land—refer to
Tasmanian Parks and Wildlife
Service.

FACILITIES

No public facilities.

SOUTHERN

OCEAN

Scale 0 1 2 3 4 5 Km

158° 57'

North Head

HASSELBOROUGH BAY

Wireless Hill

Handspike Point

The Isthmus

54° 30'

ANARE STATION

BUCKLES BAY

Scobie lake

Halfway Hill

Eagle Point

Nuggets Point
The Nuggets

Mt Elder
371

BAUER BAY

Mawson Point

SANDY BAY

Cormorant Point

Square Lake

Brothers Point

Prion Lake

Aurora Point

Mt Ifould
374

Green Gorge

Mt Waite
422

Mt Blake
371

WATERFALL BAY

Cape Toutcher

Major Lake

Mt Hamilton
433

Mt Aurora
383

LUSITANIA BAY

Mt Jeffryes
399

54° 45'

Mt Kidson
382

Caroline Point

CAROLINE COVE

Mt Ainsworth
363

South West Point

SOUTH EAST BAY

Hurd Point

ST FRANCIS ISLAND – NUYTS ARCHIPELAGO

Bight Baptism

It was not an auspicious landing. After a day-long voyage aboard a chartered fishing boat we were eager to set foot on St Francis Island, the largest of the outer islands in the Nuyts Archipelago, some 50 kilometres south of Ceduna. We came, ostensibly, in the name of science, to check on the rare creatures that reside on this castaway island. For my part I was also on the trail of a 270-year-old literary mystery.

Wedged in a South Australian Parks and Wildlife Service dinghy laden with supplies and scientific equipment, our party of four motored into Petrel Cove just on dusk. It was a calm evening with a gentle swell curling into the bay around North Point. Ahead, the slender beach backed by high dunes looked inviting. But as the dinghy nosed into the sand, two large waves suddenly reared at our backs and broke over the stern. Our gear was drenched and the boat awash. We stepped ashore shivering. Welcome to St Francis Island.

This baptism was a reminder that, even in the most placid conditions, these waters should not be underestimated. Over the years South Australia's west coast has acquired a reputation for lively storms, White Pointer Sharks (*Carcharodon carcharias*) and fearsome seas that claim the lives of local fishermen with grim regularity.

For millennia the Southern Ocean swell, urged on by strong westerlies, has eroded away the soft underside of the continent. During the last ice age, when a large expanse of the continental shelf was exposed, massive deposits of lime-rich sand accumulated over the existing bedrock and solidified to form a capping of calcarenite. The sweep of coastline known as the Great Australian Bight is testimony to the appetite of the elements for this form of limestone. These same forces have also left in their wake some 19 islands that form the Nuyts Archipelago.

In The Footsteps Of Gulliver

From the air these rarely visited outliers can appear unprepossessing. Mostly arid, low-lying and sparsely vegetated, they hardly conform to the ideal of a romantic hideaway. But first impressions are deceptive and closer inspection reveals a more complex and intriguing picture.

By an act of providence these islands have served as stationary arks for a diversity of life, including animals hounded to the brink of extinction on the mainland. The nearby Franklin Islands are, for instance, home to the world's last known natural population of the Greater Stick-nest Rat (*Leporillus conditor*), an industrious and appealing indigenous rodent once widespread across southern Australia.

In another quirk of history these islands can also claim one of the earliest European contacts with Australia. In January 1627 Peter Nuyts from the Dutch East India Company explored these waters aboard the *Gulden Zeepaard* under the command of Francis Thijssen. Apart from the naming of two of the largest islands in the archipelago, St Pieter and St Francis, after the two men's respective patron saints, little is known about the remarkable voyage of these navigators. The only surviving record of their discoveries is a 1644 chart.

For nearly 200 years these islands stood at the frontier of knowledge about the southern coastline of Terra Australis. In yet another curious twist of history, early charts provided one of the locales for what is arguably the most enduring and ingenious work of satire in the English language.

Thus, while others in my party came equipped with specially designed animal traps, measuring apparatus and vats of liquid nitrogen for freezing biological specimens, I was armed with a pocket edition of what purports to be an account of the travels of Captain Lemuel Gulliver. Jonathan Swift's *Gulliver's Travels* is a malevolent satire on the follies of eighteenth-century English society, government and law, masked as a kind of fairy tale. The peripatetic

Opposite: Aeolian calcarenite cliffs, the hardened relics of wind-borne dunes, are conspicuous on the exposed southern shores of St Francis Island. Such fretted ramparts, overlying foundations of ancient granites, give South Australian islands their distinctive countenance.
Peter Canty

Below: St Francis Island is home to the only known island population of Carpet Pythons in South Australia.
Quentin Chester

Gulliver suffers moral indignation at every landfall—Lilliput, Brobdignag, Laputa and Lagado.

Of particular interest to me was the Captain's final and perhaps most fantastic voyage of all, to a land presided over by the esteemed Houyhnhnms, those most civilised of horses. By my calculations St Francis was in close proximity to this still to be re-discovered island realm. And I hoped to find some evidence of the vile and detested Yahoos, the brutish human-like creatures that ultimately drove Gulliver to recoil in horror from his fellow man.

It was a quest I approached with some trepidation, especially in the light of the question posed by Gulliver himself: "For who can read of the virtues I have mentioned in the glorious Houyhnhnms, without being ashamed of his own vices, when he considers himself as the reasoning, governing animal of his country?"[1]

Treasured Islands

After a cool night the morning sun was a welcome addition to our campsite high on the dunes above Petrel Cove. From my tent flap I could see the length of the beach, where on 2 January 1802 Matthew Flinders aboard HMS *Investigator* landed during his detailed survey of the archipelago he named in honour of Peter Nuyts. There was no sign of Gulliver's strange beasts on the shore but in the soft, gilded light I did spy a group of promenading birds gazing out to sea. Then more of these birds appeared, this time flying in formation, each with the unmistakable profile of the Cape Barren Goose (*Cereopsis novaehollandiae*).

During the winter months these large grey birds are the most conspicuous tenants on St Francis, one of many such breeding islands in southern Australian waters. Trekking to the island's lighthouse, I encountered many more geese, some on nests, others chomping on fresh winter grass. Though handsome to look at, their incessant honking and squabbling over territory is decidedly Yahoo-like. It seemed hard to believe that they were close to extinction in the 1950s. And yet, despite their

*Above: Quandong (*Santalum acuminatum*) and Coast Bonefruit (*Threlkeldia diffusa*) on St Peter Island. Peter Canty*

Right: Once prevalent across southern Australia, the sole surviving population of Greater Stick-nest Rats was discovered on the Franklin Islands in 1920. Concerted efforts over the past two decades have seen additional populations established on St Peter Island and Reevesby Island in the Sir Joseph Banks Group, in an effort to ensure the survival of this endangered and endearing native rodent. David Carter

spectacular recovery, they remain one of the world's rarest species of goose.

The lighthouse stands on the rounded 81-metre-high summit at the south-eastern end of St Francis. Attractive native shrubs including *Westringia* sp., *Templetonia* sp. and *Correa* sp. dot the slopes nearby. But most of the island's 809 hectares are covered in saltbush and undulating grassland with occasional sandhills and scattered limestone outcrops. To the northwest, the island drops away to Petrel Cove. A freshening wind brought cloud and fine, misting rain from across the sea. In the distance stood two abandoned farmhouses, a forlorn reminder of the years around the turn of the century when a succession of pastoralists leased St Francis for sheep-grazing.

Sitting atop this remote windswept isle, it seemed to me a brave notion that such a place could sustain viable agriculture. But the two long-term lessees, Thomas Lloyd and William Arnold, showed remarkable enterprise. As well as running sheep, they grew crops of lucerne, wheat and barley, as well as assorted vegetables. They kept chickens, ducks and turkeys. As a sideline they even mined and exported penguin guano. But by 1939 the effects of droughts, over-grazing and isolation put an end to agriculture on St Francis. In 1972 it was dedicated as a conservation park.[2]

In the wake of these decades of disturbance

it is surprising that there was anything much left to conserve. The most insidious legacy of this era is the African Boxthorn (*Lycium ferocissimum*), introduced by the farmers for windbreaks and hedges. Armed with sharp spikes and fleshy leaves, it is a widespread, intractable presence, competing with native species and colonising sea bird nesting sites. A plant only a Yahoo could love.

However, apart from this intrusion and the eradication of Brush-tailed Bettongs (*Bettongia penicillata*) by the farmers' cats, most of the native animals survived. This makes St Francis, along with the other islands in the archipelago, a priceless asset for science and conservation. Over the past decade biologists from the National Parks and Wildlife Service have been closely studying the wildlife of these islands and working to ensure that species like the Greater Stick-nest Rat have an assured future, both in the archipelago and in sanctuaries elsewhere. It is a redemptive mission worthy of the wise Houyhnhnms.

Of special interest to our scientific party were the Southern Brown Bandicoots (*Isodon obesulus*) on St Francis, one of only two such populations of a species now threatened on the mainland.[3] Despite setting some 80 traps overnight not a single Bandicoot was snared by this method. Perhaps these intelligent-looking animals had become preternaturally wary of small aluminium boxes bearing peanut butter and oats. In the end, it was only by using more rudimentary techniques, such as flushing individuals out of large bushes with our bare hands, that any specimens were obtained.

We found that many of these bushes also harboured a variety of reptiles. St Francis has various geckos, six species of skink and the spectacular and venomous Black Tiger Snake

(*Notechis scutatus*), a renowned predator of lizards and small birds. Even more noteworthy are the resident Carpet Pythons (*Morelia spilota*). This is the State's only known island refuge of these majestic snakes with their distinctive yellow, gold and black markings.

As one who has not made a habit of being intimate with cold-blooded creatures there was something strangely uplifting about handling these pythons. Aside from their elegant patterning and graceful power it was their placid temperaments that impressed me the most. And if cranky Gulliver could apprehend profound truth and wisdom from a strange race of horses, why couldn't I find a little wonderment in a noble snake?

Pursuit At Trivia Bay

In my remaining time on St Francis I headed cross-country to the rugged cliff-tops overlooking Trivia Bay. It was tricky walking. The sandy loam covering this stretch of the island is riddled with burrows, the deserted dwellings of Short-tailed Shearwaters (*Puffinus tenuirostris*), some 270,000 of which breed here each summer. One false step sees you plunging up to the knee into subterranean nests. Adding to the embarrassment of the moment is the knowledge that these burrows are also the preferred haunts of the Black Tiger Snakes.

When Matthew Flinders and company visited St Francis it was high summer and "the great

*Pride of St Francis. The sandy bays and granite pavements of the Nuyts Archipelago are among the most significant basking and breeding sites for the Australian Sea-lion, one of the world's rarest seal species.
Peter Canty*

Hunted in the early 19th century for their oil and hides, Australian Sea-lions are now restricted to the southern shores of South Australia and Western Australia, with an estimated total population of 9,000 to 12,000 animals. It is thought the gestation period for Australian Sea-lions is up to 18 months—some 7 or 8 months longer than other seal species.

heat" and lack of water, "even to rince our mouths", curtailed their interest in the island. But the shearwaters were in residence and Flinders noted, "I found the surface of the island where it was sandy and produced small shrubs, to be full of their burrows." So despite the island's barren aspect, shore parties gathered several hundred shearwaters to distribute among the crew. According to Flinders the birds would be considered by most people inedible "on account of their fishy taste; but they made a very acceptable supply to men who had been many months confined to an allowance of salt meat."[4]

The sound of the Southern Ocean crashing into the bay was audible long before I reached the coast. When I finally made it to a suitable vantage point, the reverberations of the breakers were as deafening as artillery fire. In contrast to the tranquillity of Petrel Cove, where we had watched Bottlenose Dolphins (*Tursiops truncatus*) frolic and Little Penguins (*Eudyptula minor*) waddling ashore after a day's fishing, the fierce countenance of this aspect of the island was anything but trivial. It was not hard to imagine the reefs below being patrolled by White Pointers and brutal seas thundering against this exposed shoreline.

I continued along the edge of the brittle limestone cliffs but, despite an extensive search, I still found no trace of the savage and sublime creatures that Gulliver described in such graphic detail. The literary mystery remained. But at the same time I felt closer to understanding what had disturbed him so. Often it takes a journey to a place like St Francis to be reminded of the Yahoo and Houyhnhnm within us all.

1   J. Swift, *Gulliver's Travels*, Hamish Hamilton, London, 1947, p. 324.

2   A.C. Robinson *et al*, *South Australia's Offshore Islands*, Denle, 1996.

3   P.B. Copley, V. T. Read, A.C. Robinson and C.H.S. Watts, 'Preliminary Studies of the Nuyts Archipelago Bandicoot *Isodon obesulus nauticus* on the Franklin Islands, South Australia, in J.H. Seebeck *et al*, eds, *Bandicoots and Bilbies*, Surrey Beatty & Sons, Sydney, 1990, pp. 345-56.

4   M. Flinders, *A Voyage to Terra Australis*, G. & W. Nicol, London, 1814, vol. 1, p. 116.

ST FRANCIS ISLAND – NUYTS ARCHIPELAGO

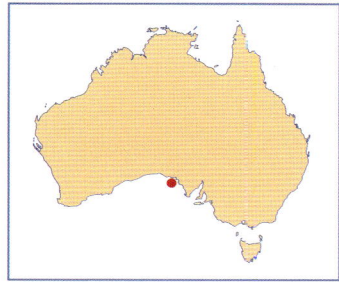

LOCATION

32° 30' S, 133° 17' E. Approximately 40 km south-west of Ceduna.

AREA

St Francis: 809 ha
St Peter: 4,028 ha
Franklin: 405 ha

CLIMATE

Temperate/Mediterranean

STATUS

Nuyts Archipelago Conservation Park, managed by the South Australian National Parks and Wildlife Service.

ACCESS

Private vessel or charter boat from Ceduna. No public access to Franklin Island without written authorisation.

FACILITIES

None

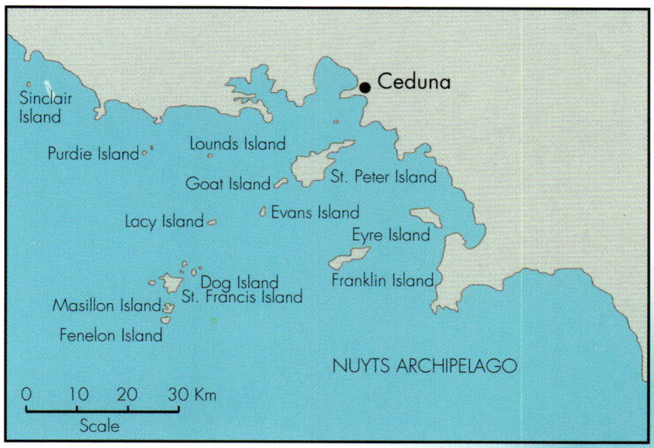

Ceduna
Sinclair Island
Purdie Island
Lounds Island
Goat Island
St. Peter Island
Evans Island
Lacy Island
Eyre Island
Franklin Island
Dog Island
St. Francis Island
Masillon Island
Fenelon Island
NUYTS ARCHIPELAGO
0 10 20 30 Km
Scale

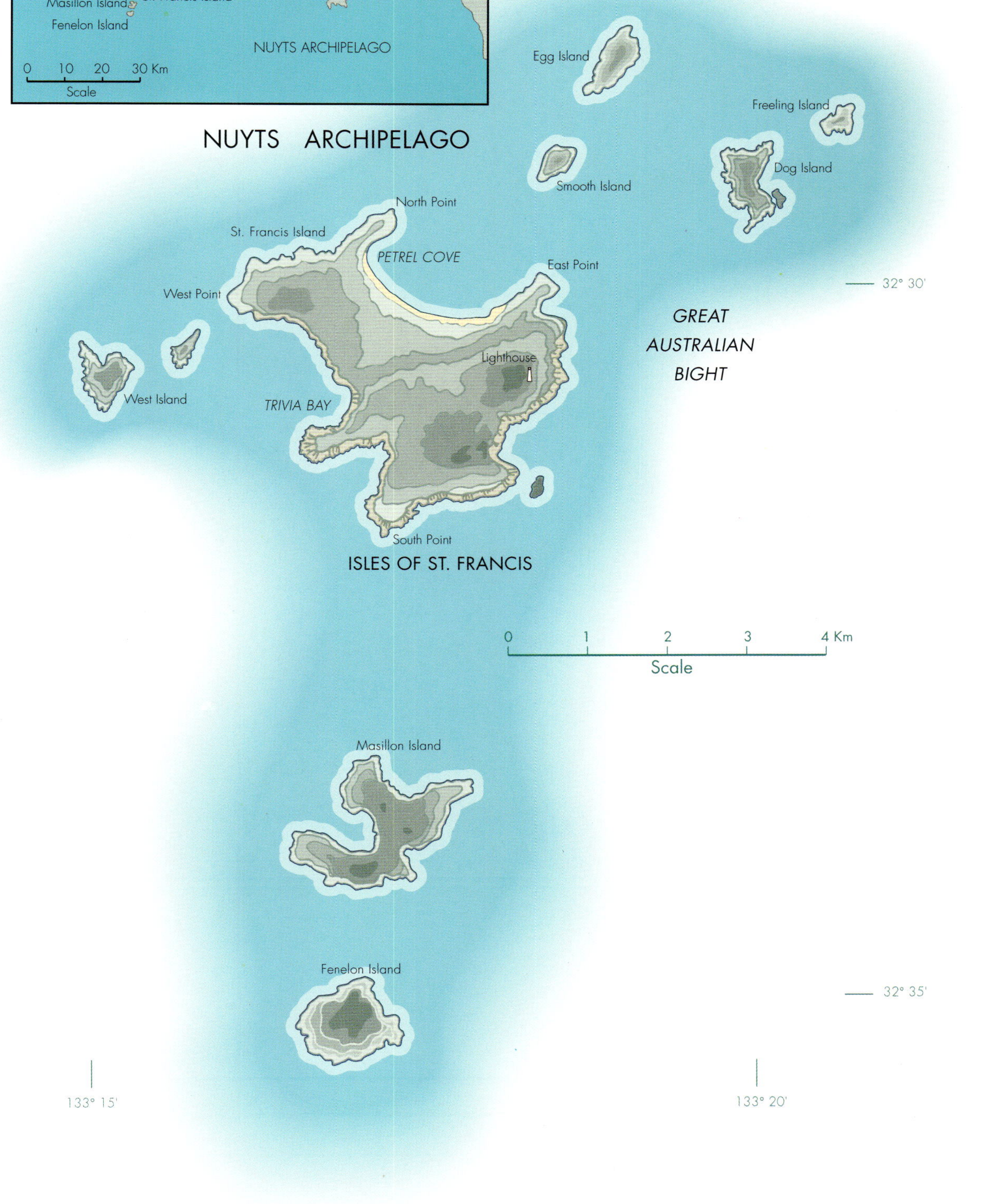

NUYTS ARCHIPELAGO
Egg Island
Freeling Island
Dog Island
Smooth Island
North Point
St. Francis Island
PETREL COVE
East Point
West Point
GREAT AUSTRALIAN BIGHT
Lighthouse
West Island
TRIVIA BAY
South Point
ISLES OF ST. FRANCIS
32° 30'
0 1 2 3 4 Km
Scale
Masillon Island
Fenelon Island
32° 35'
133° 15'
133° 20'

MONDRAIN ISLAND – ARCHIPELAGO OF THE RECHERCHE

Bearings

At 345 metres Mount Le Grand is the highest point in a 200-kilometre stretch of Western Australia's southern coastline. The true summit of this granite dome is a flat expanse of rock fringed by shoulder-high scrub. The view inland is extensive but the coast remains partially obscured by vegetation. Only by bustling through the thickets of banksias and eucalypts to rocks standing proud on the southern spur of the mountain is it possible to observe the complete panorama.

With compass at the ready, I gazed out from this vantage point into the setting sun and a brisk south-westerly breeze. Some 260° of the view was taken up by white caps scuffing the Southern Ocean swell. Scattered across this sweep of water, from the shore to the horizon, was a great constellation of islands. Some rose from the sea in rotund or conical peaks similar in shape to Mount Le Grand. Other distant examples lay long and low in the water as vague silhouettes. When waves struck these sleek profiles and sent spouts of spray into the air, the islands took on the appearance of a distant pod of whales surfacing in the twilight.

While the map flapped wildly in my hands I identified some 50 or so individual islands before the light faded and the wind made my eyes stream. The impact of this vista was underscored by the knowledge that my tally represented barely half the total number of islands in the archipelago. Similarly, the smaller rocks and exposed reefs I could see, encircled by creaming white surf, were only a small fraction of the 1,500 islets scattered along this span of coast.

Fleeting Glimpses

The Archipelago of the Recherche is one of the most extensive congregations of islands visible from the Australian mainland. Dispersed across some 13,000 square kilometres of open sea, the group includes some of the least visited and most undisturbed islands in the country. While several are within sight of the coast and some are connected by sandspits to the mainland at low tide, the outermost members of the archipelago—Termination and Salisbury islands—lie 50 kilometres offshore, at the farthest limits of the continental shelf.

Salisbury is regarded as Western Australia's oldest continental island, having been separated from the continent by rising sea levels some 14,000 years ago. While there is archaeological evidence of Aboriginal occupation among the islands, research suggests that human activity pre-dated their separation from the mainland. It is thought that the majority of the offshore islands have remained free from human activity for at least 8,000 years.[1]

The first recorded European sighting of the archipelago was made by Pieter Nuyts during the remarkable voyage of the Dutch ship the *Gulden Zeepard* in 1627. An interval of 164 years passed before George Vancouver aboard the *Discovery* brushed past the southern edge of the group in 1791 and named Termination Island. Then, in the following year, Bruni d'Entre-casteaux, in command of *La Recherche* and *L'Espérance*, visited the group during his voyage in search of the missing expedition of François de La Pérouse.

Retreating from south-westerly gales, the French ships took shelter among the islands. As well as naming the archipelago and several islands, d'Entrecasteaux landed on one of the inshore islands.[2] Meanwhile, the naturalist Jacques-Julien Houten de Labillardière gathered plant specimens, one of the earliest scientific collections of Western Australian flora. But the greater part of the archipelago was left unexplored—a challenge taken up by Matthew Flinders a decade later.

Forbidding Shores

The most conspicuous island visible from Mount Le Grand is Mondrain. Lying 10 kilometres to the south-west, it was named by d'Entrecasteaux after the midshipman of *La Recherche*, Pierre Mondrain. It is the second-largest island in the archipelago. From rounded northern headlands it extends six kilometres south along a ridge to a broad, conical-shaped peak rising 222 metres. While Mondrain's central elevations are clad in a patchy mantle of vegetation, its shoreline is dominated by slabs of granite. On those shoulders of the island exposed to storms and fierce southerly winds, these slabs reach to within a few metres of the island's summit.

Forbidding shorelines are a hallmark of the archipelago. Only three islands—Figure of Eight, Middle and Sandy Hook—have anything resembling a sand beach. The majority of the group are encircled by wave-swept aprons of stone. These perimeter defences have been a key factor in keeping the islands relatively free from outside disturbance. The perils of landing dissuade most casual visitors and, as a consequence, the impact of feral animals and introduced plants has been negligible. For those determined to make landfall—most often, it seems, scientists studying the natural

history of the group—the essential equipment is a wet suit and a shrewd eye for the ups and downs of the Southern Ocean swell.

On larger islands like Mondrain there is, however, often a lee shore that offers some respite from the prevailing winds, if not complete protection from the surge of the swell. The conditions for our landing on Mondrain Island were exceptionally benign—so much so that the abalone boat that delivered us to the island was able to edge to within jumping distance of the shore on the island's south-western flank.

For weeks I had been steeling myself to swim through foaming seas. I read accounts by John Béchervaise, the leader of a pioneering scientific expedition to the archipelago in 1950. He described "broad surf-polished ramps covered in goose barnacles and treacherous black slime" and the experience of "clawing at cracks and scrabbling out of reach of the surge".[3] My disembarkation, by contrast, entailed stepping from the boat's gently bobbing bow onto a ramp of warm, dry stone.

Fields Of Colour

From a distance the domed flanks of Mondrain gave the illusion of being smooth, uniform expanses. But traversing the slabs on foot gives a very different impression. The rock's colours, textures and weathered formations are constantly changing. The crystalline texture of the rock gives good friction. By keeping one's ankles canted, so that sole rubber is flat on the rock for maximum grip, it's possible to sidle up steep slopes in the manner of a crampon-shod mountaineer ascending hard-packed snow.

While the geology so evident on Mondrain, and throughout the archipelago, is often called granite, the predominant rock type is migmatite. Dating from Pre-Cambrian times, more than 600 million years ago, it is composed of lath granite and garnet gneiss. Dark bands of biotite and amphibolite are common and on Mondrain there are extensive dykes rich in quartz and felspar. Its diverse mineral content gives the rock a mottled appearance in shades of pink, grey, yellow and brown. These inherent colours are embellished by tigerish stripes of

Like starbursts, the flowers of the Bushy Yate and its urchin-like woody fruits are an arresting sight in woodland communities on the islands of the archipelago.
Rob Jung

algal black and ever-present lichens in masses of sage green and vivid orange. To walk these slabs is to trek across fields of colour, a palette of tonal abstraction on a colossal scale.

In contrast to the jumble of dense scrub at higher elevations, Mondrain's exposed faces provide ready access to other parts of the island. On the island's bare southern slopes, sustained weathering has created a knobbed surface of indentations, some like monstrous eye-sockets. Crossing these pockmarked expanses is akin to roaming a mountain of recently cooled magma.

This lunar landscape is the domain of the Ornate Rock-dragon (*Ctenophorus ornatus*). Mondrain is one of the few known island strongholds of this small dragon lizard. They are ubiquitous on the island's warm granite faces and their territorial perching and antic displays of head-bobbing were a constant source of diversion as we traipsed across the slabs. When approached, the lizards would eventually abandon their defiant postures and sprint to safety under rocks and exfoliated flakes.

At sea level the granite slabs become smooth, washed by the incessant wave action. Walking north along the western flank of the island we skirted close to the water's edge. These margins were patrolled by a pair of Sooty Oystercatchers (*Haematopus fuliginosus*) who prodded and probed for morsels exposed by the receding tide. As we came near they tooted anxiously at us in their faintly comic manner, "like wooden toys with whistles", to quote Béchervaise's apt description.

Rock Dwellers

Half-way along the island the clean line of the sloping shore gives way to a succession of inlets, where the sea has quarried away at joints and weaknesses in the rock. Here we scrambled across headlands and wove our way through huge rock outcrops, perched high above wave-cut platforms. There, among the overhangs and dark fissures, we glimpsed the wallabies of Mondrain for the first time.

For an hour we played a game of hide and seek with these rock-dwellers. Though wary and

watchful of our presence they were also highly inquisitive. As we edged nearer, they appeared to look away for a possible exit and then turn back, tilting their striped heads as if to see us better, seemingly caught between the impulse to take flight and a desire to scrutinise their visitors at close range.

Climbing to the top of an outcrop, we startled one of the group. This individual retreated onto a sheer-sided buttress, which seemed to offer no avenue of escape. The animal stood transfixed. Then, in an explosive instant, it turned and launched itself into the air, flying in a daring arc across a deep void, before landing on an adjacent buttress and bounding out of sight.

Since 1905 the Black-footed Rock-wallabies

Harried by persistent south-westerly winds, the vegetation of Mondrain Island's exposed faces strains to keep its footing on the rocky pavements.
Rob Jung

that inhabit the Archipelago have been considered a subspecies, *Petrogale lateralis hacketti*. Commonly known as the Recherche or Hackett's Rock-wallaby, these compact, beautifully marked animals are found on Mondrain, Wilson and Westall islands. Meanwhile, Middle and North Twin Peak islands are the preserve of the Tammar Wallaby (*Macropus eugenii*). The precise reasons for the distribution of the two species remain unclear. It's likely, however, that in the time since the islands have been separated from the mainland, competition for suitable habitat and other, random factors have resulted in one or the other species gaining ascendancy.

Charting The Coast

The first record of the Recherche Rock-wallaby was made by Matthew Flinders in the early stages of his epic circumnavigation of the continent. On the evening of 8 January 1802 his ship, HMS *Investigator*, had approached the westerly islands in the group. Flinders wrote, "The French Admiral [d'Entrecasteaux] had mostly skirted round the archipelago, a sufficient reason for me to attempt passing through the middle . . . if the weather did not make the experiment too dangerous."[4] The resolve of the young captain was tested late the following day when his ship stood off Mondrain Island with no shelter at hand.

Continuing our trek to the northern point of the island we crossed low headlands to overlook a strait of exposed water extending to the mainland. Pacific Gulls wheeled aloft in a buffeting easterly breeze that ripped across the sea. To the east stood a cluster of islands and to the west lay reefs and rocky shoals washed by thunderous white breakers. The hazards of this unknown coast for a cumbersome vessel reliant on sail power alone were alarmingly apparent.

With no safe passage to clear water, Flinders took the bold decision to run for the mainland. There, sanctuary was found in a sandy cove he named Lucky Bay. From this anchorage forays were made onto the adjacent shore and out to Mondrain. Lieutenant Fowler and Mr Thistle returned from the island with "some seals of a reddish fur, and a few small kangaroos of a species different from any I had before seen."

In the course of their visit the island was set alight and Flinders notes that "the wind being fresh, there was a general blaze in the evening all over the island." This is the first record of fire in the archipelago. While lightning strikes would have always caused a number of naturally occurring blazes, it's likely that, with the increasing European presence in the years to follow, many more islands would have been ravaged by fire.

Salisbury Island, with its limestone capping, has the potential to support a particularly formidable vegetation cover. For James Willis, the botanist on the 1950 Béchervaise Expedition, this scrub was the most impenetrable he had ever encountered. "The bushes grew in unbelievable density, a solid thicket of interlacing branches up to shoulder level . . ."[5] But then a lightning strike in 1992 saw fire sweep over the island, wiping out much of the vegetation and starting another cycle of re-growth.

Lawless Desperados

Matthew Flinders spent 10 days surveying the archipelago. In the manner of the times islands were charted and named, botanical specimens collected and wildlife both duly noted and eagerly consumed. Geese, assorted fish, seals and presumably rock-wallabies helped supplement the ship's rations. Of the seals Flinders wrote: "All the islands seem to be more or less frequented by seals; but not I think in numbers sufficient to make speculation from Europe advisable on their account."

This opinion did not, however, deter the eventual drift of sealers into the archipelago from their strongholds on Kangaroo Island and the islands of Bass Strait. Seal colonies were plundered mercilessly in the quest for pelts. Encampments on the more accessible islands became staging posts for a tenuous, frontier existence.

Major Edmund Lockyer, who arrived in 1827 to establish a settlement at King George Sound, was confronted by a band of desperate sealers whose supply ship had failed to arrive. He comments in his journal that they were

Opposite: Monarchs of Mondrain. Ever-inquisitive Black-footed Rock-wallabies are a beguiling presence among the lower granite outcrops of the island. Forming localised colonies of thirty individuals or more, they inhabit fissures and overhanging boulders, feeding on vegetated terraces nearby.
Rob Jung

Overleaf: From the granite crown of the southern summit, the view north along Mondrain's rounded flanks reveals evening mist drifting past the mainland peaks of Cape Le Grand on the mainland.
Rob Jung

Supremely agile in broken terrain, the Black-footed Rock-wallaby has granulated soles on its feet, giving exceptional grip on rock, while a slender tail provides crucial balance for aerial manoeuvres.
Rob Jung

"obliged to live on anything they could get, even a dog; they have with them one Hundred Fur Seal Skins and have about Seven Hundred on an island near Mondrain Island."[6]

Another, later account describes Middle Island being "the resort of a set of lawless desperados, composed of runaway convicts, sealers etc."[7] The most infamous of these was an American, 'Black Anderson'. According to a report in The Perth Gazette of 7 October 1842:

Anderson usually carried a brace of pistols about him, knowing that he held his life by a very precarious tenure. By persevering exertions he had amassed a considerable sum of money, and usually kept one or two black women to attend on him and minister to his wants, when not engaged in sealing.

While the moral state of the sealers attracted much speculation there was also official concern about the future of the seals. As early as 1827 Edmund Lockyer wrote, "if some measures are not immediately resorted to, [the sealing trade] must be irreparably injured if not destroyed altogether."

Despite such calls, it would be nearly a century before seals gained any form of protection. Given the wholesale slaughter it is surprising that extermination was avoided. In the years since they have been protected, both species have recovered significantly and are widely distributed among the islands, though the total populations are still below the levels that prevailed prior to commercial hunting. According to one recent count, the archipelago is now home to nearly 1800 New Zealand Fur Seals (*Arctocephalus forsteri*) and approximately 1200 of the relatively rare Australian Sea-lions (*Neophoca cinerea*).[8]

Troubled Waters

The Archipelago of the Recherche has been a nature reserve since 1948. While seals and other wildlife have generally prospered, the area is not immune from external threats. This was graphically illustrated on 14 February 1991 when the 33,000-tonne *Sanko Harvest* ran aground on what was later called Harvest Reef, some 12 nautical miles south-west of Esperance. By the following Saturday the ship began to break up and its 30,000-tonne cargo of phosphate and 800 tonnes of bunker oil spilled freely from the wreck.

Over the following weeks a large-scale clean-up operation was carried out, involving 30 personnel from the Department of Conservation and Land Management (CALM), together with more than 100 volunteers. Thanks to their efforts the impact of the oil spill was contained. Nevertheless some 15 seal pups and some 30 seabirds died. And the effects of the slick on intertidal marine life in the period immediately following the accident were devastating.[9]

The combination of lively seas and smooth granite shorelines did keep most of the islands relatively free from any long-term build-up of oil residues. The mainland shores were, however, less fortunate. As we had discovered on a walk along the coast from our lookout on Mount Le Grand, the high water mark on some sheltered coves and tidal pools is daubed with a black scum that looks and smells like tar. Six years after the spill the surface of the water in these pools is still streaked with an oily film. Meanwhile the *Sanko Harvest* now rests some 26 metres below the surface and is steadily being appropriated by fish and other marine creatures as an artificial reef. It is also a lure to scuba enthusiasts, being the second-largest vessel in the world that can be dived on.

From Mondrain's southern summit, Harvest Reef lies to the west, lost among the other islands and shoals, just another hazard in what Flinders described as "this extensive mass of dangers". Looking out across the water one sees no sign of shipwrecks. Nor does Mondrain yield any evidence of the sealing days, or obvious scars left by the fires lit by Flinders's men—nor indeed any of the other more recent blazes. Instead, for the unwitting bystander the illusion is of an island—an archipelago—inviolate.

On Top

To reach Mondrain's southern summit we had followed a spiral path winding anticlockwise up the dome of granite, past exfoliated flakes and shards. Across this shedding scalp of a mountain stood islands of vegetation where plants had taken hold in seepage cracks and the dark, sandy humus of small terraces. Assorted cushion plants and noon flowers clung defiantly to these perches. Other isolated woody shrubs were pruned into long, flattened shapes by the prevailing winds. Most dramatic were the dead limbs of a prostrate Tea tree (*Leptospermum* sp.), sculpted by the elements into grotesque scrolls of bleached wood.

Higher on the lee side of the summit we entered mixed forests of She-oaks (*Casuarina* spp.) and Island Paperbarks (*Melaleuca globifera*). The Bushy Yate (*Eucalyptus cornuta*) was the dominant eucalypt, a species with clusters of strange finger-like buds and large woody fruit. Where the forest canopy was thick, the interlocking branches and fallen timber were swaddled in pastel-shaded lichens and dripping with primeval strands of mistletoe. By contrast, a few metres away were sheltered clearings with dwarf conifers, cushion plants, mosses and trigger flowers, bordering small rockpools with smooth pebbles of granite— vignettes as ordered and serene as any Japanese courtyard.

For several nights we camped close to the granite outcrop that capped the summit. With their curved, under-cut faces and dark cavities

The expansive rocky slopes of Mondrain Island are populated by the Ornate Rock-dragon. Like many dragon lizards they employ a range of display gestures.
Rob Jung

these blocks took on unnerving forms, appearing at times like the discarded molars of some huge mastodon. At other moments they served as viewing platforms. From these eyries we observed the mercurial moods of the archipelago as warm winds blew from assorted points of the compass.

We shared the summit with Welcome Swallows (*Hirundo neoxena*) who seemed to delight in darting from their nest in the cool crevices of the outcrop. Not far away several hundred Fleshy-footed Shearwaters (*Puffinus carneipes*) crowded each night into burrows on a grassy south-facing slope, their summer tenancy of the island having just begun, their eggs about to emerge for incubation. Another neighbour with more than a passing interest in these eggs was a large Carpet Python (*Morelia spilota imbricata*) that each morning draped itself across the patio of dark rock alongside our camp to luxuriate in the unseasonal heat.

Boom And Bust

Just below our camp were shallow pools in rock depressions. But the impression they gave was misleading. In truth it had been a lean year. Many parts of the island we visited were tinder-dry with some shrubs and meadows appearing stressed by a lack of water. While much of the island fauna is adapted to cope with drought periods, the survival of one species, the Recherche Cape Barren Goose (*Cereopsis novaehollandiae grisea*), is precarious in such conditions.

Around 75 islands in the archipelago are used as nesting grounds for these large grey birds whose predominant diet consists of green grasses. While the populations of Cape Barren Geese in Bass Strait and South Australia number some 12,000 and 3,000 birds respectively, the Recherche subspecies is much less common, with an estimated population of no more than 600 to 1,000 birds.

Successive hot, dry seasons can have a severe impact on island life, killing off vegetation, including the exposed grassy clearings which the geese rely on for their grazing. Some surveys conducted during drought years have suggested that the total population has plummeted to as few as 200 birds, pitching the subspecies into the endangered category, albeit temporarily. Follow-up census work indicates that the species has the potential to recover from these events but the dynamics of these population shifts are not yet fully understood.[10]

By our final day on the island the small pools near the summit had completely evaporated. The ephemeral plants dependent on this moisture were beginning to shrivel and desiccate. With the onset of summer it could be many months before run-off from fleeting showers would collect in these depressions again—and perhaps much longer for the population of Recherche Cape Barren Geese to reassert its vigour.

These islands may be well defended but they are far from impregnable or unchanging. Their salvation remains their isolation, from one another, as well as from human disturbance and the mainland. It is an isolation worth defending and deserving of further study. But no amount of research will contain the daunting spectacle of all these isles, nor fully fathom the secrets of their seasons and cycles. To that extent the islands of the archipelago remain as wondrous in their mystery as the greatest of all mammals that migrate through these waters—the whales they can so strikingly resemble.

1   A. Burbidge, *Western Australian Islands*, CALM

2   Several sources cite Observatory Island as the one in question, though others believe it was Woody Island.

3   J. Bechervaise, 'The Archipelago of the Recherche', *Walkabout*, April, 1951.

4   M. Flinders, *A Voyage To Terra Australis*, facsimile edition, S.A. Govt Printer, 1989.

5   J. H. Willis, 'Among Plants of the Recherche', *Walkabout*, August, 1951.

6   Edmund Lockyer, 11 March 1827, Journal.

7   *Inquirer* (Perth), 5 January 1848.

8   A. Burbidge, B. Haberley, S. Halse, J. Lane, G. Pearson, 'How Many Geese are Enough?', *Landscope*, 9 (1), pp. 127–33.

9   A. Storrie, G. Pobar, 'Harvest from the Sanko', *Landscope*.

10   A. Burbidge *et al, op cit.*

MONDRAIN ISLAND – ARCHIPELAGO OF THE RECHERCHE

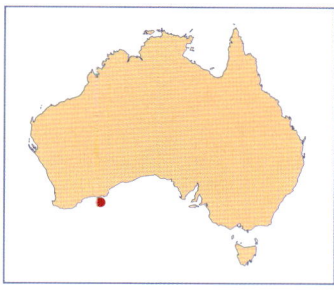

LOCATION

Mondrain Island 34°07' S 122°15' E , 40 km south-east of Esperance. The archipelago extends for some 200 km, from 121°30' E to 124°30' E. It includes 105 islands and 1500 islets.

AREA

810 hectares

CLIMATE

Temperate

STATUS

Mondrain is part of the Archipelago of the Recherche Nature Reserve, managed by the Western Australian Department of Conservation and Land Management (CALM).

ACCESS

Private vessel or charter boat from Esperance. Regular sightseeing cruises visit the islands of Esperance Bay.

FACILITIES

None on Mondrain. Camping facilities available on Woody Island.

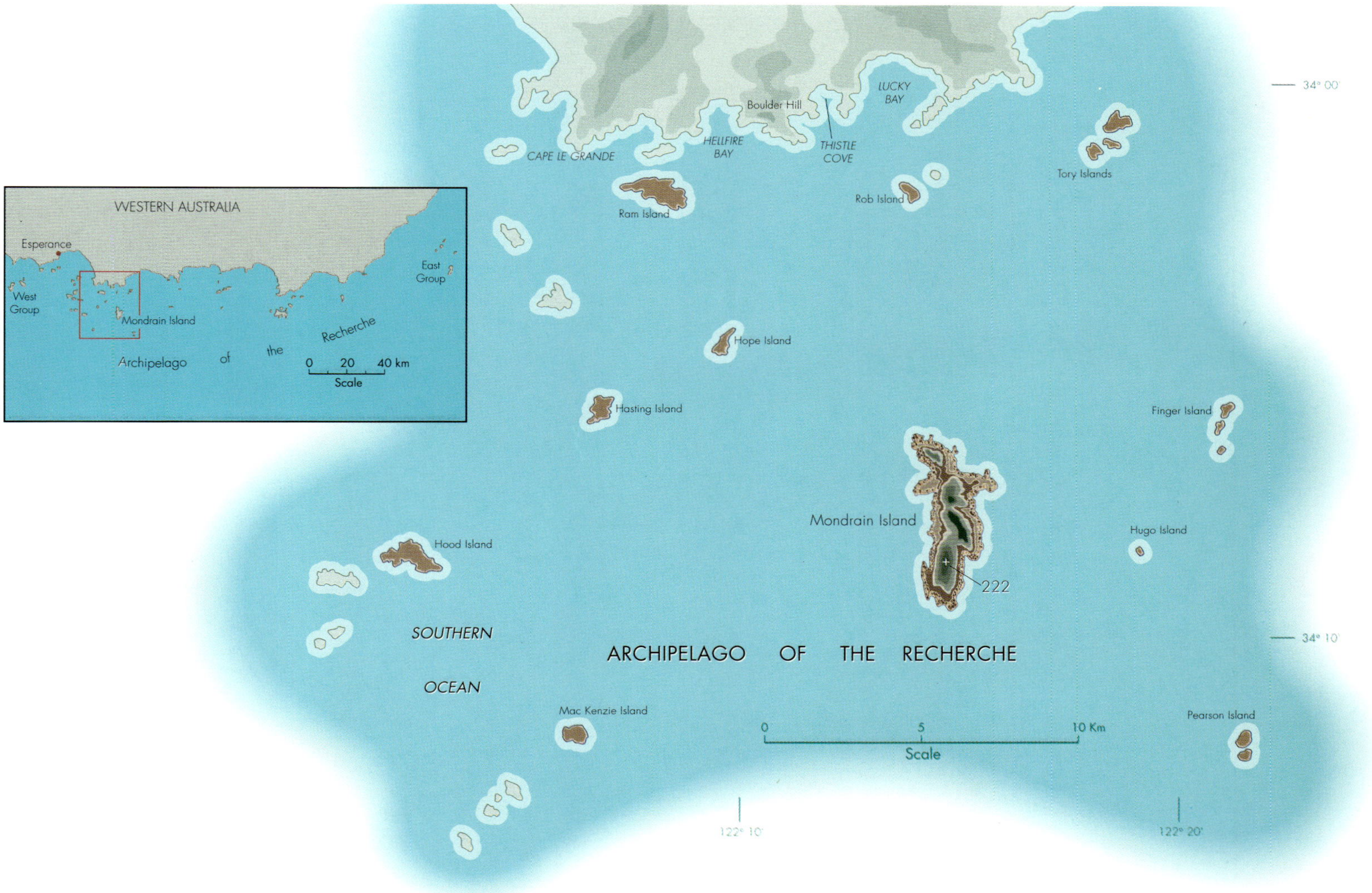

WESTERN AUSTRALIA

Esperance

West Group

Mondrain Island

East Group

Archipelago of the Recherche

0   20   40 km
Scale

34° 00'

Boulder Hill

LUCKY BAY

CAPE LE GRANDE

HELLFIRE BAY

THISTLE COVE

Tory Islands

Ram Island

Rob Island

Hope Island

Hasting Island

Finger Island

Mondrain Island

Hugo Island

222

Hood Island

SOUTHERN

OCEAN

ARCHIPELAGO   OF   THE   RECHERCHE

34° 10'

Mac Kenzie Island

Pearson Island

0   5   10 Km
Scale

122° 10'

122° 20'

Glacier's Retreat (Brown Lagoon, Heard Island)
Oil on canvas 122 x 107cm

FIRE AND ICE

HEARD ISLAND

*In profound isolation, the ice-clad flanks of
Australia's only active volcano rise in
defiance against the raging Southern Ocean.
Despite the assaults of near constant gales,
Heard Island's shores are a haven for vast
congregations of wildlife.*

HEARD ISLAND

Grand Isolation

That moment had been a long time coming. After nearly four weeks at sea, pounding under sail through the justly named Roaring Forties and Howling Fifties, nothing could replace the first sight of those islands of grand isolation.

The French territory of Îles Kerguelen, our only port of call since leaving Fremantle, was nearly two days astern to the north-west. Within the space of a few hours, the already chill wind had become markedly colder. Sleet replaced rain. We had crossed the unseen line known as the Antarctic Convergence, the northern limit of waters trapped in near continuous circulation back and forth from Antarctica. With a sudden drop in sea temperatures of 2° Celsius, this amorphous yet clear divide separates cold-temperate oceans from icy polar seas.

We careered onwards down the face of awesome swells; sails, sheets and spars straining, racing before wind and spray. Dozens of albatrosses and petrels revelled in the conditions, gliding with indifference, their incisive wings cutting the air just centimetres above the wave-tops. It was a typical summer's afternoon in the Southern Ocean.

Our first sight of land arrived like an apparition in the nothingness of ocean. Meyer Rock, an outlying pinnacle of the McDonald Group, thrust its grey-black finger from the waves. McDonald Island itself then hove into view. Assaulted by enormous waves, we cautiously left it well to starboard. Largest in a cluster of three small fragments of land, 60 kilometres west of Heard Island and one of the loneliest places on Earth, only once has a successful landing been made on McDonald from the sea.

Before any details of its bulk could be deciphered, we were conscious of an odour that would soon become very familiar. The two kilometre-long McDonald Island is a place of seabirds, in particular Macaroni Penguins (*Eudyptes chrysolophus*), crowding its precarious heights in an aggregation approaching one million pairs.

Rushing onwards into the last watery streaks of drawn-out summer light, we at last saw Heard Island's unwelcoming countenance. Through sharp squalls and flurries of sleet we nosed into Atlas Roads and a tenuous anchorage under the cliffs of the Laurens Peninsula at the western end of the island.

We had arrived at just about the most isolated place imaginable. Anywhere was distant; everywhere was distant: 4,100 kilometres north-east to the coast of Western Australia, a similar distance to South Africa and still 1,700 kilometres to Antarctica. Heard Island is Australia's most remote sovereign territory.

Wind, Ice and Fire

Forty-two kilometres from end to end and 20 kilometres across at its widest, Heard Island is vaguely reptilian in shape with a head, grossly distended body and flimsy tail. Its enormous bulk is formed by the mountain known as Big Ben. Bound in snow and ice, glaciers grind a short but dramatic path from the flanks of Big Ben to the coast — more than 2,000 metres in just a few kilometres. Several of the glaciers discharge directly into the sea, ending abruptly in ice-cliffs up to 50 metres high.

From behind the semi-circular remnant wall of an extinct volcanic caldera, rises the active cone of Mawson Peak, a further 700 metres above the summit plateau's snow bowl. At 2,745 metres, Mawson Peak atop Big Ben has the distinction of being Australia's highest mountain outside Antarctica and its only active volcano. Lava is known to have last spilled over its lip in 1992.[1]

The island's climate is severe. Cold, squally and cloudy weather is common. It lies directly in the path of the westerly wind-drift in a region where a near-continuous stream of cold fronts and cyclonic outbreaks develop. Average winter

Opposite: The Compton Glacier and the North Barrier. Most Heard Island glaciers have been in rapid retreat over the last decade or so, with the greatest change occurring to those at the eastern end of the island. Since this photograph was taken in the early 1980s the snout of the Compton Glacier has retreated almost a kilometre further inland.

"Slots on the Baudissin Glacier, 1948"—looking towards the Laurens Peninsula. From 1947 to 1955 Heard Island served as a proving ground for personnel, sledging dogs and equipment prior to the establishment of a permanent Australian presence on the Antarctic continent.
G. Budd. Courtesy of Australian Antarctic Division, Hobart

discovered in the region. The hunt for whales and seals probably led to an unrecorded chance encounter in an infinity of ocean but the truth of its discovery has always been mysterious. James Cook, exploring with the *Resolution* and *Adventure*, recorded signs of land at the island's approximate position in February 1773. However, the first sighting was most likely made in 1833 by the British whaling captain and explorer Peter Kemp as he ranged south from Îles Kerguelen in search of seals. In command of the *Magnet*, Kemp reached the coast of Antarctica only to be washed overboard on the return journey. Back in England his logs also disappeared but a form too large to be an iceberg was noted on the chart lodged with the British Admiralty.[2]

Twenty years later, in November 1853, the island was seen by American merchant Captain John Heard sailing the Great Circle route from Boston to Melbourne aboard the barque *Oriental.* His was the first definitive sighting, and the island duly received its name. A few months later a group of small islands to the west were seen by Captain William McDonald, master of the British sealing vessel *Samarang.*

Soon after, Heard Island almost claimed its first shipwreck. The *Earl of Eglinton,* carrying nearly 400 migrants from Scotland to Australia came close to disaster off the Laurens Peninsula before dawn on 1 December 1854.

"Desolation Island"

While Heard and the McDonald Islands were slow to appear on charts in general use, commercial interest developed quickly. In 1854 whaling and sealing operations in the Southern Ocean were well established, with gangs having worked the Falkland Islands as early as 1775 and South Georgia from 1786. More than 70 years' experience foretold that the beaches would be littered with corpulent Southern Elephant Seals (*Mirounga leonina*). Whale and seal oil were in great demand for lighting, lubricants and general industrial purposes. Antarctic Fur Seals (*Arctocephalus gazella*), prized for their pelts, were also to be expected.

Two years after Heard's sighting, the slaughter commenced. American sealers from

temperatures at sea level are close to freezing, rising by 4° Celsius in summer. Snow can fall at any time on the coast and often accumulates in deep drifts during winter. Rain, sleet or snow have been recorded on an average of 300 days in a year and gales gusting faster than 200 kilometres per hour have been known to tear across the island. Fierce winds drain cold air off Big Ben, the massive bulk of the mountain making for violently unpredictable conditions and pronounced microclimatic effects.

Signs of Land

Secreted deep in the Southern Ocean, Heard Island was the last significant land mass to be

New London, Connecticut, were first when Captain Erasmus Darwin Rodgers and a party from the *Corinthian* landed in January 1855. Rodgers sent word home of great riches.

Another ship, the *Laurens*, was hastily dispatched, joining the *Corinthian* and its five tenders — exploring the island, naming its features and butchering its wildlife. In the year immediately following the first landing, nearly 3,500 barrels of elephant oil and 500 fur seal pelts were shipped away.[3] The cargo of the *Laurens* alone was valued at $130,000. Soon, separate gangs were wintering over. Each worked its own beach or stretch of coastline, whipping, clubbing, and shooting these defenceless creatures, their great bulk making them easy victims where they lay. Thick strips of blubber — the seal's fatty layer of insulation — were flensed from each carcass, rendered down to oil by boiling in large iron cauldrons known as trypots, then decanted into barrels ready for collection by returning ships. Alternatively blubber was taken directly aboard waiting vessels and processed at anchor or back at Îles Kerguelen. After three years 25,000 barrels of oil had been shipped back to the United States. This represented the slaughter of approximately 8,000 Elephant Seals.

Throughout the sealing days American interests predominated but their early monopoly was short-lived. The whalers of Hobart soon learned of the wealth of Heard Island. W. L. Crowther, owner of one of the largest fleets, dispatched ships in 1858. Others were lured south from the Cape Colony in South Africa. At the height of activity between 1858 and 1862 more than 100 men worked the beaches.

The sealers went about their work in conditions of great hardship. They were marooned with few provision on beaches flanked by impassable tongues of ice, uncertain of when or if their ship would return. One of the first Australians ashore gave a vivid account of their privations:

"We were all ordered on shore on an island covered with ice and snow and without any shelter or covering for our heads or bodies and amid the most intense and bitter cold with snow dropping, having no fuel to make a fire ... At length we succeeded in finding along the beach some old portions of a wreck with which we managed to erect a place sufficiently large for us to crawl into ... in this place, you will hardly credit it, we had to exist for upwards of six months ... our fireplace was tussocks of grass and our fuel consisted of elephant blubber and penguin skins, for we could get no other".[4]

Another observer described life on the island as "mist, snow and slaughter, the packing of oil, hard bread and bad beef, fatigue and heavy slumbers..."[5] Just as the sealers preyed on the wildlife, so the isolation, hardship and danger preyed on their minds, proving too much for a few. Madness shrieked in the relentless gale as some men summarily fell to bullets otherwise intended for seals. At least nine ships foundered on the island's coast, dashed against glaciers and treacherous promontories. Deaths were many. Valuable cargoes, hard-won through the sufferings of beast and man, often failed to reach a waiting ship, let alone the lamps of Sydney, London or New York. Instead, they poured from capsized barges and smashed barrels, mingling with the icy seas, swallowed up in an instant by a maelstrom of surf. Heard Island became known to the sealers as "Desolation Island".[6]

With no thought to maintenance of the resource, the seal populations of Heard Island were brought to near-extinction within a few years. King Penguins (*Aptenodytes patagonicus*) were also sacrificed for oil, resulting in temporary extinction that lasted until the 1960s. Voyages became rare as profits dwindled. The 1880 wreck of the New London ship *Trinity* effectively marked the end of the sealing voyages to Heard Island.

During a short revival, Norwegian and South African interests made a handful of visits in the early decades of this century. On one such occasion in 1910 the British flag was hoisted at Atlas Cove.

In the following years science replaced profit as the motivation for a few brief sojourns before the first Australian National Antarctic Research Expedition (ANARE) set up a permanent base in December, 1947. Sovereignty simultaneously passed to Australia. As a proving ground for men and equipment prior to the establishment of an Australian presence in Antarctica, the

Heard Island Station at Atlas Cove was maintained until 1955. It was then given up in favour of Mawson Station on the MacRobertson Land coast, 1,700 kilometres to the south. Once abandoned, Atlas Cove was quickly reclaimed for nature, buildings broken apart by gales and the lumbering weight of Elephant Seals. That short period of less than eight years was the first and last time Heard Island has been continuously occupied since the sealing era. Subsequent visits have been infrequent and mostly brief. After 1955, the only party to have wintered over, spent 14 months at Spit Bay, commencing in the summer of 1992.

Over a century has passed since sailing ships regularly anchored in Atlas Roads. The curses of the men that were brought to Heard Island; the dreadful isolation and the raw, turbulent majesty of that sea-mountain of ice and fire are today the attraction of a truly wild place. We also came under sail, on the well-known 27-metre maxi-yacht *Anaconda*, comparatively safe and in much greater comfort, trusting in a sturdy ketch and her seasoned skipper, the redoubtable Josko Grubic.

Island Building

In the cool southern seas of the middle Eocene to early Ogliocence (52 to 24 million years ago) a broad limestone plateau was laid down stretching south-east from the current vicinity of Îles Kerguelen to about 62° South. Deposited atop extensive extrusions of 115 million year-old basalt, the Kerguelen Plateau rises over 3,000 metres from the surrounding ocean floor to within a kilometre of the sea surface. Commencing with massive movements 10.5 to five million years ago, the limestone basement of the island, brimming with tiny marine fossils, was thrust upwards. It can still be seen as the underlying strata on stretches of the Laurens Peninsula coastline. However, all this was but a prelude to island building on a grand scale.

Between five and three million years ago underwater volcanism increased dramatically. As an island emerged, a jumble of volcanic material from numerous explosive events interbedded with sediments, forming agglomerates — something akin to a "geological fruit cake". Erosion of the new-born island was balanced by the frequent intrusion and outpouring of basalt as agglomerates continued to form — evident today on the Laurens Peninsula in strata nearly 400 metres thick.[7] Marine deposits from this time provide fossil evidence of warmer seas and an island clothed in scrubby vegetation.

After yet more erosion, island building reached its finale. Volcanic eruptions took place on a massive scale as highly viscous basaltic lava spewed from vents somewhere about the centre of the island. Big Ben, a classic conical volcano, accumulated rapidly. Separate volcanic activity on the Laurens Peninsula saw Anzac Peak and Mount Olsen grow. The most recent phase of volcanism produced Mawson Peak and some spectacular coastal features, including Rodgers and Corinth Heads, and Mount Dixon, a volcanic cone of trachyte, basalt, lava, ash and cinders that now stands more than 700 metres above sea level. Coinciding with successive global ice ages, glaciation began in the Pleistocene Epoch.

Cabbage and Cushions

About 80 per cent of the island is bound in snow and ice. Of the ice-free area, much is bare rock, sand and rubble. To date, only 11 species of vascular plant, 42 species of moss and less than 50 lichens have been recorded. Among the vascular species, *Poa cookii*, *Azorella selago* and Kerguelen Cabbage (*Pringlea antiscorbutica*) are most widespread.

Cushions of *Azorella* are found from the wettest of coastal areas to the most exposed limits of plant colonisation, among high windswept stretches of feldmark. On a broad wedge of coastal flat near Dovers Moraine in the east and also at Atlas Cove, *Azorella* mingles with *Poa* tussocks to form a waterlogged labyrinth of small freshwater pools. North of the ANARE Station and exposed to the full blast of the wind, a pure stand of *Azorella* interspersed with colonising tufts of moss, carpets the appropriately named Azorella Peninsula. In a broad velvety expanse, acid-green *Azorella* cushions make a dazzling, almost gaudy, contrast with the gunmetal lava to which

Above: Brooding King Penguin, Spit Bay. Kings, and the larger Antarctic-breeding Emperor Penguins (Aptenodytes forsteri), do not build nests. Instead, they incubate the eggs and cosset their new-born chick cradled on their feet, all covered by a brood pouch.

Right: Macaroni Penguins on Capsize Beach. The most numerous bird species breeding on Heard Island, Macaronis congregate in enormous colonies on steep slopes and rocky promontories. The Heard Island population may exceed one million breeding pairs, with a further million pairs found on the McDonald Islands. Breeding between October and January, Macaronis lay two eggs but only one chick normally survives.

they cling so tenaciously.

Kerguelen Cabbage is equally striking. Leathery cabbage-like rosettes and prominent yellow-green fruiting clusters add a distinctive flamboyance to much of the coastal landscape. In well protected locations it grows with startling fecundity.

Southern Ocean Serengeti

After a day spent installing two members of our party in one of the few habitable buildings remaining of the ANARE Station, we ventured away from the dubious safety of Atlas Cove. Motoring out around the Azorella Peninsula then along the north coast towards Spit Bay, we were graced with an afternoon of rare calm.

Close inshore, *Anaconda* loped past the great black battlements of Rodgers and Corinth Heads and then a sweep of glaciers. The Baudissin, Challenger, Downes and Ealey — angry, fractured snouts of ice — each gouged its way out of the clouds and into the sea. Ice gave way to the steely dark flanks of 400 metre-high Round Hill, its lower slopes streaked in green. For the next six weeks, Round Hill was

to be a familiar landmark for 12 of our land party camped on a stretch of bare ground at Skua Beach between the Brown and Stephenson lagoons.

After the most anticlimactic of landings, with almost no surf or wind to contend with, we "dug in", securing our tents behind makeshift protective walls of beach shingle. A dose of Heard Island's more typical weather was sure to be on its way.

A mountaineering assault on Big Ben was the centrepiece and ultimately the major achievement of the expedition. But as the island is so rarely visited, we involved ourselves in an array of field tasks for research scientists back in Australia with interests as diverse as glaciology, botany and zoology.[8]

Away from Skua Beach, I made many trips in nearly all weathers along a stretch of coast from the outfall of the Compton Glacier to the Elephant Spit, Winston Lagoon and Cape Lockyer. Each time, I was privileged to watch at close quarters the process of life unfolding, exactly as it had for unknown generations. Like an oceanic Serengeti, the scarce habitable parts of the coast were crowded with life. Apart from its resident seals, the island is an important breeding station for 19 bird species, with huge populations among Macaroni Penguins, Rockhopper Penguins (*Eudyptes chrysocome*) and South Georgia Diving-petrels (*Pelecanoides georgicus*). Burrowing petrels and prions, and vulnerable ground nesting birds such as the island's small population of exquisite Antarctic Terns (*Sterna vittata*) are free to breed and raise their young away from predation by cats and rats. The island is also a stopover for a number of travellers, itinerants and vagrants. Chinstrap Penguins (*Pygoscelis antarctica*) come ashore and the most travelled of all birds, the Arctic Tern (*Sterna paradisaea*) lives a life of perpetual summers alternating between Heard Island and its breeding grounds in the Arctic, nearly 20,000 kilometres away. The island is also a winter resort for that denizen of the Antarctic pack ice, the Leopard Seal (*Hydrurga leptonyx*), which often hauls up on the beaches around Atlas Cove.

Unlike all other major islands of the region, Heard remains just about free of exotic plants and animals. During the sealing era everything from reindeer to rabbits were released on the islands of the Southern Ocean. The intent of their liberators was to provide food for sealers and marooned sailors. Rats escaped of their own accord from innumerable shipwrecks, while cats were taken ashore as pets and to control rats. It is quite probable that both found their way to Heard Island but none have persisted, possibly due to the cold.[9] The island's ecosystem has survived in its pristine state. The only naturalised alien plant, *Poa annua*, a northern hemisphere annual meadow-grass, is thought to be a recent arrival. A coloniser of disturbed ground, it was perhaps spread to Heard Island by seabirds.[10]

Kings and Regiments

Spit Bay was the scene of much activity in the sealing era. The coastal plain east from our camp was scattered with poignant reminders of the past. Lines of broken barrels, still holding solidified seal oil, stood like tombstones on the beachhead, while rusting trypots filled by wind blown sand were potted-out with causal arrangements of Kerguelen Cabbage and

tussock grass.

The remains of a sealers' stone and sod shelter near the lagoon at Spit Bay was surrounded by an incongruous juxtaposition of life. Several thousand King Penguins had re-established a colony within snorting distance of a large wallow of moulting Elephant Seals. About 50 seals lay in a densely packed, steaming pile amid an expanse of mud and denuded tussock mounds. Two creatures could not have looked more unlikely together. Beside tonnes of loafing leviathan and the malodorous mix of penguin and seal filth, the splendour of those imperious birds seemed somehow more intense.

The King Penguins in the Spit Bay colony were closely but evenly spaced, each defending its own standing room from its neighbour with a swipe of a bill or the slap of its flipper. Most birds were incubating eggs or brooding newly born chicks.

Although Elephant Seals have recovered well from the predation of last century to a total population of about 85,000 in the mid 1980s, there has been an as yet unexplained decline in the number of pups born each season since 1948. Running at about 40 per cent, it is consistent with similar declines on other sub-Antarctic islands. This contrasts with the exponential rise in Antarctic Fur Seal numbers on Heard Island and an equally dramatic increase in sub-Antarctic Fur Seals (*Arctocephalus tropicalis*) on many islands in the Southern Ocean.[11]

Kings interspersed with Gentoo Penguins

*Now regarded as an endemic species, Heard Island Cormorants (*Phalacrocorax nivalis*) number fewer than 100 pairs. These rare birds, along with other species, may be vulnerable to any disturbance of marine ecosystems in the region. In 1996 a limited commercial trawling licence was issued for the Australian Fishing Zone (12 to 200 nautical miles) around the island. A similar fishery has operated around Macquarie Island for the past few years. All such activities need to be carefully monitored.*

Above: Young male Elephant Seals spar in shallow waters near the outlet to the Brown Lagoon. At the end of winter, mature males often engage in bloody fights for control of a stretch of beach and the "harems" of pregnant cows. Dominant bulls are known as beachmasters.

Opposite: Fruiting rosettes of Kerguelen Cabbage, a widespread and common colonising plant found in all of the island's six plant communities. Its tightly enfolded fleshy leaves are a haven for wingless flies. In a devolutionary adaptation to an extremely windy environment, all of the island's insects are flightless.

Right: Elephant Seals showing signs of the annual summer moult. From November to February, they frequent extensive mud wallows among coastal tussocks, intent on easing the discomfort of shedding old skin.

Riviera Days

It was a fine afternoon, so I stepped out along the sands of the Spit and on into the wilds of the Southern Ocean, almost walking on the sea. The Spit had grown in the lee of the island, pushed eastward for nine kilometres by wind and furious surf. Even on a calm day, the wash on each side left me feeling quite vulnerable, as if this umbilicus of sand was about to be cut through, leaving nothing to the north, south and east in a 270° arc but thousands of kilometres of open ocean.

Running short of time and having walked only about two kilometres along its length, I reluctantly turned for the long trudge back to camp. The majestic isolation of the island of ice ahead, lifted my step and my spirit. The entire face of Big Ben from the North Barrier and the Compton Glacier to the South Barrier and Lambeth Bluff was free of cloud. In the enduring afternoon light, its flanks were suffused with a delicate crystalline blue. Clearly defined ribs of rock buttressed the thousand metre-high flanks of the caldera. The Compton, Brown, Stephenson and Winston Glaciers fanned from its base in a nearly unbroken sweep of ice, their slopes rippling with crevasses and teetering séracs. For the Stephenson Glacier, it was a steep and rough path down, abruptly halted short of the sea by

(*Pygoscelis papua*) marshalled in scattered groups of 50 or 100 birds out from the breeding colony, across the long grey arc of the island's alluvial tail and on into the far distance. Resplendent in their plumage they resembled soldiers arrayed on the battlefield in showy eighteenth-century uniforms.

the blue-grey band of Dovers Moraine. A jumble of stale ice and rubble merging with vegetated slopes, the moraine was strewn across the landscape like overburden stripped from some enormous excavation

Aside from rapid fluctuations in the capricious and violent weather, the very physical nature of the place is still undergoing great change. In 1987, a summer ANARE party noted that the Spit had been breached, leaving Spit Point a stranded sandbar, separated from the body of the island by a stretch of turbulent shoal water. It has since reformed and split again.

Similarly but on a grander scale, if I had stood on the Spit and observed Heard Island's glaciers just 30 years earlier, significant differences would have been apparent. Air and sea temperatures in the region have risen by as much as 1° Celsius over the past 50 years, leading to rapid changes in the island's ice cover.

In 1859, the Compton Glacier's smaller neighbour, the Brown, claimed one of the island's early victims when the *Mary Powell*, carrying a cargo of 400 barrels of oil, ran under its ice cliffs and was wrecked. Today, all that a foundering ship might strike at the same point would be loafing Elephant Seals, hauled out on the bar fronting a recently formed lagoon. Likewise, the ice in Dovers Moraine is melting as the Stephenson Glacier withdraws, allowing wild surf and storms to tear at the base of the island's tail.

Stopping for a moment near Scholes Lagoon at the island end of the Spit, I looked down from the monumentality of the mountain to the details at my feet. The hardened crust of sand was embedded with pewter-grey lumps of basalt, faceted and worn smooth by gales and the blast of sand. Like weather vanes ground in steel-hard stone, they told of the fury that is so often Heard Island.

Continuing on to camp, it seemed hard to reconcile the signs of such violence with the glories of that calm summer's evening. Big Ben hung upside-down in a glassy Stephenson Lagoon, reflections of pink cloud tufts brushing grounded shards of ice. Gentoo Penguins, those ubiquitous beach-front residents, stood passively among the forward

tussocks, each gilt-edged with light, while in the shallows gleaming Antarctic Fur Seal pups frolicked and splashed.

In the lee of the mountain, the north-east end of the island experiences brief intervals in the weather stream that border on the delightful. Despite times of wind, rain and driving snow, I often felt we had come to the Riviera of the Southern Ocean with gentle days of cool breezes, easy curling waves, and as ever that mountain rising from the sea into a cobalt sky. But such a false sense of ease is unwise in such an unpredictable place.

Loneliness Reclaimed

As if we were being told to leave, the island soon turned, wrecking our puny camp in one short episode with winds gusting well past 100 kilometres per hour. We had conquered the mountain and our presence was no longer tolerated. Returning with two companions from Winston Lagoon a few days before, we had been repeatedly blown off our feet and drenched in gale-driven spray while crossing Dovers Moraine, more than a kilometre from the sea.

It was almost March when we left Heard Island for an exhilarating run back to Adelaide. Winter was not far off and the island had reclaimed its loneliness with an emphatic measure of force. Trying to summarise the experience, I commented in my diary shortly after weighing anchor: "Remarkable beauty, either startling or subtle from beginning to end is how I will always remember it ..." Contentment and privilege, I also sensed: contentment that we had managed to weather that island of wind, ice and fire almost as the sealers had; and privilege in being able to visit such a well-preserved and unique environment. To me, Heard was a place of spiritual rather than commercial enrichment and definitely not "Desolation Island". What a difference a century makes.

1   During our visit steaming, sulphurous fissures and vents were examined on the summit of Mawson Peak. They have now given way to a 100 metre wide by 60 metre deep cauldron of molten lava.

2   Cook made meticulous observations on all his voyages. During those of the *Resolution* and *Adventure* he noted seals and whales in abundance in the vicinity of South Georgia and the South Sandwich Islands, triggering great interest in the entire region. His signs of land in the vicinity of Heard and the McDonald Islands would not have gone unnoticed.

    On 13 February 1773 Cook noted: "We were now accompanied by a very much greater number of Penguins than at any time before ... the meeting with so many of these Birds gave us still hopes of meeting with land ..." J. C. Beaglehole (ed.), *The Journals of Captain James Cook On His Voyages of Discovery*, vol. 2, Hakluyt Society, Cambridge, 1961, p. 93.

    Because of its proximity to Îles Kerguelen, the scene of much whaling and sealing activity in the early 1800s, Heard Island may have received secretive visits from fur sealers around that time. P. D. Shaughnessy, G. L. Shaughnessy and P. L. Keage, 'Fur Seals at Heard Island: Recovery from past exploitation?' in M. Augee (ed.), *Marine Mammals of Australasia*, Royal Zoological Society of New South Wales, Sydney, 1988, pp. 71–7.

3   The small numbers of fur seal pelts taken in the first recorded years of sealing compared to recovered population levels in recent years adds weight to the conclusion that secret visits were made to the island before 1855. Nearly 9,800 seals were recorded in a February 1988 census. *Ibid.*, pp. 71–7. See also S. D. Goldsworthy and P. D. Shaughnessy, 'Counts of the Antarctic Fur Seal', *Arctocephalus Gazella* and 'Locations of Colonies at Heard Island in the 1987–88 Summer',

ANARE Research Notes 72.

4   Quoted in P. G. Law and T. Burstall, *ANARE Interim Reports* 7, Heard Island, Antarctic Division, Melbourne, 1953, p. 3.

5   *Ibid.*, p.2.

6   A name Heard Island shared with Îles Kerguelen. P. Keage, 'The conservation status of Heard and McDonald Islands', *University of Tasmania Environmental Studies Occasional Paper* 13, 1981.

7   The strata is known as the Drygalski Formation. Baron Erich von Drygalski, leader of the 1902 German Antarctic Expedition called briefly at Heard Island aboard the *Gauss* and made the first comprehensive scientific observations.

8   Several unsuccessful attempts at climbing Big Ben were made during the ANARE days, 1947–55. The summit was first reached in 1965 by a party of five led by mountaineer and adventurer Warwick Deacock. It was Deacock's second attempt. Their climb began at the western end of Winston Lagoon. See P. Temple, *The Sea and the Snow: the South Indian Ocean Expedition to Heard Island*, Cassell, Melbourne, 1966.

9   Heard Island's climate is similar to that of South Georgia, where cats and rats failed to survive. A. E. Burger, 'Terrestrial Food Webs in the Sub-antarctic: Island Effects' in W. R. Seigfield *et. al.*, *Antarctic Nutrient Cycles and Food Webs*, Springer-Verlag, Berlin, 1985.

10 J. J. Scott, 'New Records of Vascular Plants from Heard Island', *Polar Record* 25 (152).

11 This species was first recorded breeding on Heard Island in 1987. *Heard Island Wilderness Reserve Management Plan*, Australian Antarctic Division, Hobart, 1996.

HEARD ISLAND

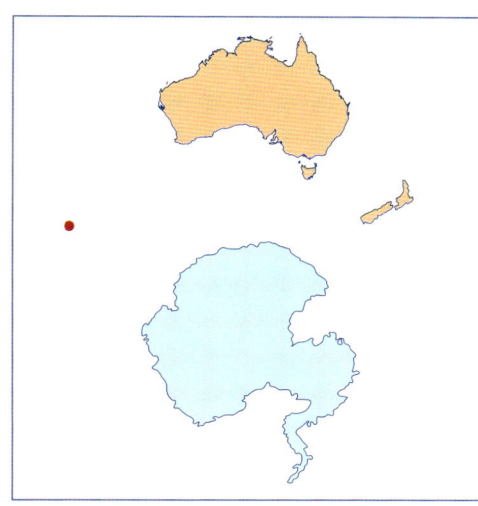

LOCATION

53°05' S, 73°30' E.
(McDonald Islands: 53°03'
S, 72°35' E). 4,100 km
south-west of Fremantle;
similar distance from Cape
Town, South Africa.

AREA

Heard Island: 36,800 ha
McDonald Island: 100 ha

CLIMATE

Severe—cold, wet, windy

STATUS

Australian External
Territory, nominated for
World Heritage List in
1996. Designated as Heard
Island Wilderness Reserve
by Management Plan, 1995.
Administered by the
Australian Antarctic
Division.

ACCESS

Intermittent visits by
Antarctic Division ships.
Occasional private visits.
Access can be dangerous

and landing is weather
dependent. Permit
required—refer to the
Antarctic Division.

FACILITIES

None

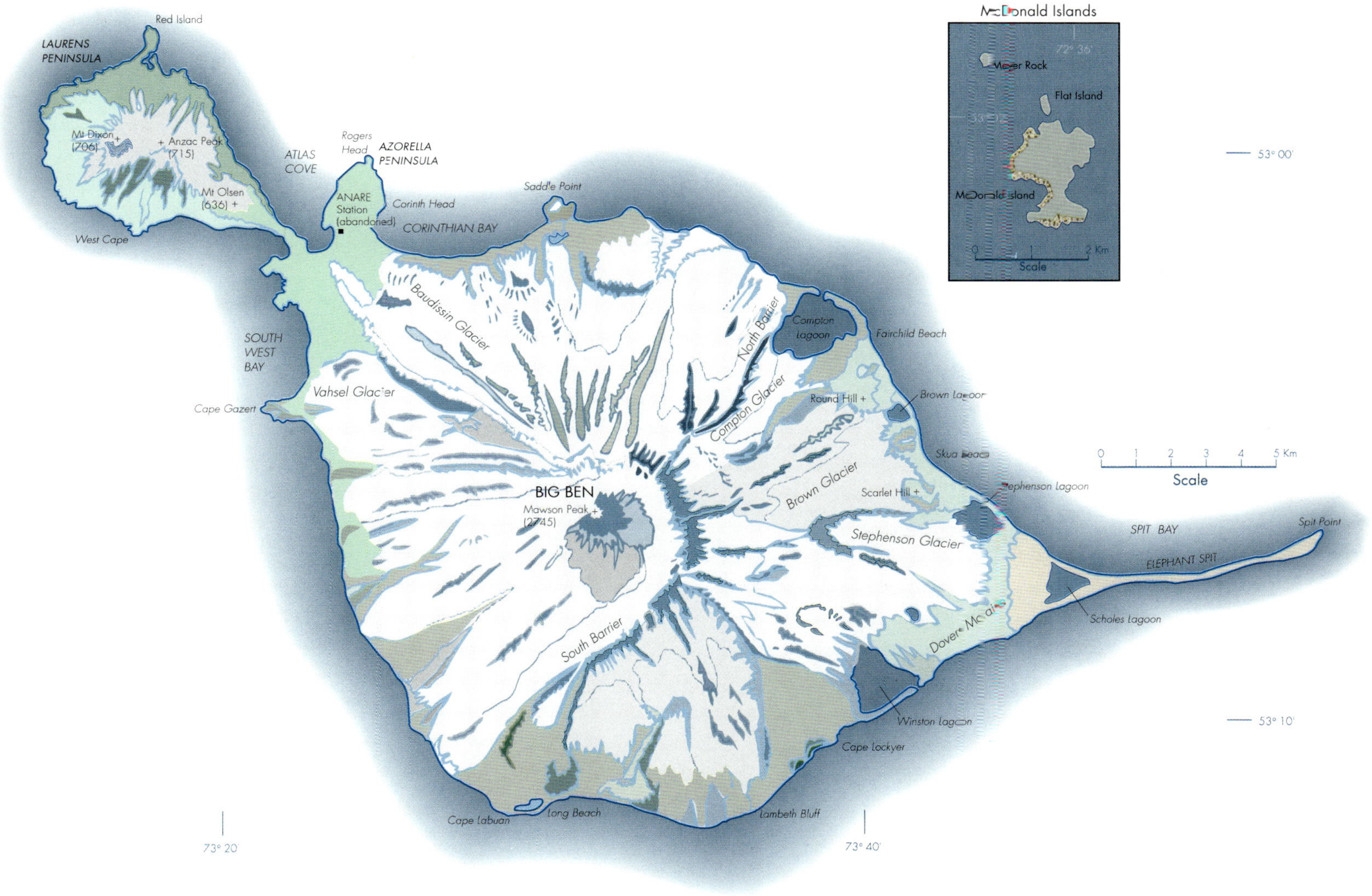

BIBLIOGRAPHY AND FURTHER READING

Authors note: With the exception of historical references, this list includes only those items not already included in chapter footnotes.

GENERAL

Baglin, D., and Mullins, B., *Islands of Australia*, Ure Smith, Sydney, 1972.

Banfield, E.J., *The Confessions of a Beachcomber* [1908], Viking O'Neil, Melbourne, 1974.

Carson, R., *The Sea*, Paladin, London, 1991.

Dutton, G., (ed), *The Book of Australian Islands*, Macmillan, Melbourne, 1986.

Hamilton-Paterson, J., *Seven-Tenths*, Hutchinson, London, 1992.

Lucas, A., *Cruising the Coral Coast*, Horwitz Grahame, 1988.

Rolls, E., *From Forest to Sea*, University of Queensland Press, St Lucia, 1993.

NATURAL HISTORY

——— *Flora of Australia*, Vol. 49, Oceanic Islands I, Australian Government Publishing Service, Canberra, 1994.

——— *Flora of Australia*, Vol. 50, Oceanic Islands II, Australian Government Publishing Service, Canberra, 1993.

——— *Reef Notes*, (various titles) Queensland National Parks and Wildlife Service, Dept. of Environment and the Great Barrier Reef Marine Park Authority, Townsville.

Burbidge, A.A., *Importance of Australian Islands for Mammal Conservation*, unpubl. paper. Dept. of Conservation and Land Management, W.A.

Burbidge, A.A., (ed.) *Australian and New Zealand Islands: Nature Conservation Values and Management*, Proceedings of a Technical Workshop. Barrow Island, Western Australia, 1985. Occasional Paper 2/89. Department of Conservation and Land Management, Western Australia, 1989.

Carolin, R. and Clarke, P., *Beach Plants of South East Australia*, Sainty and Assoc., Sydney,1991.

Daikin, W. J., *Australian Seashores*, Angus & Robertson, Sydney, 1980.

Darwin, C., *The Voyage of the Beagle*, J.M. Dent & Sons, London, 1980.

Dorward, D. *Wild Australia*, Collins, Melbourne, 1977.

Flannery, T., *The Future Eaters*, Reed Books, Sydney, 1994.

Jones, D, and Morgan, G., *A Field Guide to Crustaceans of Australian Waters*, Reed, Sydney, 1994.

Lindsey, T., *The Seabirds of Australia*, The National Photographic Index of Australian Wildlife. Angus and Robertson, Sydney, 1986.

MacArthur, R.H., and Wilson, E.O., *Theory of Island Biogeography*, Princeton University Press, Princeton, 1967.

Mather, P. and Bennett, I., *A Coral Reef Handbook*, Surrey Beatty, Sydney, 1993.

Mitchell, A., *A Fragile Paradise*, Collins, London, 1989.

Pringle, N.D., *The Shorebirds of Australia*, The National Photographic Index of Australian Wildlife. Angus & Roberstson, Sydney 1987.

Quammen, D., *The Song of the Dodo: Island Biogeography in an Age of Extinctions*, Scribner, New York, 1996.

Strahan, R., (ed) *The Australian Museum Complete Book of Australian Mammals*, The National Photographic Index of Australian Wildlife. Collins, Angus & Roberstson, Sydney 1991.

Thompson, C., (ed) *North-West Bound - From Shark Bay To Wyndham*, Dept. of Conservation and Land Management, Perth, 1990.

Wallace, A.R., *Island Life, or the Phenomena and Causes of Insular Faunas and Floras, Including a Revision and Attempted Solution of the Problem of Geological Climates*, AMS Press Inc, New York, 1975.

Wilson, E.O., *The Diversity of Life*, Harvard University Press, Cambridge Massachusetts, 1992.

Weiner, J., *The Beak of the Finch: A Story of Evolution in Our Time*, Alfred A. Knopf, New York, 1994.

Zann, L.P., *Our Sea, Our Future*. Major findings of the State of the Marine Environment Report for Australia, Department of Environment, Sport and Territories, Canberra, 1995.

HISTORY & EXPLORATION

Allen, J. and Corris P., (eds) *The Journal of John Sweatman, a Nineteenth Century Surveying Voyage in North Australia and Torres Strait*, University of Queensland Press, St Lucia, 1977.

Baudin, N.(trans. C. Cornell), *The Journal of Post-captain Nicholas Baudin*, Libraries Board of SA, 1974.

Beaglehole, J. C.(ed), *The Journal of Captain James Cook on His Journeys of Discovery*, Hakluyt Society, Cambridge, 1955.

Dampier, W., *A New Voyage Round The World*, London, 1697. (edn., Dover, New York, 1968.)

Dampier, W., *A Voyage To New Holland*, London, 1729. (Spencer, J.,[ed.], edn., Alan Sutton, Gloucester, 1981.)

Flinders, M., *Voyage to Terra Australis*, 2 vols., Nichol, London, 1814. (facsimile edn, Library Board of South Australia, Adelaide, 1966.)

Flinders, M., *Observations on the coasts of Van Diemen's Land, on Bass Strait and its islands, and on part of the coast of New South Wales*, John Nichol, London, 1801.

Flood, J., *The Riches of Ancient Australia*, University of Queensland Press, St Lucia, 1990

Grey, G., *Journals of two expeditions of discovery in North-West Australia and Western Australia* 1837-1839, vols 1 & 2, T.& W. Boone, London, 1841. (Facsimile edn, Hesperian Press, Perth, 1983)

Headland, R. K., *Chronological List of Antarctic Expeditions and Related Historical Events*, Cambridge University Press, Cambridge, 1989.

Henderson, G., *Unfinished Voyages, Western Australian Shipwrecks 1622 - 1850*, University of Western Australia Press, Perth, 1980.

Holthouse, H., *Ships in the Coral*, Macmillan, Melbourne, 1976.

Hordern, M., *Mariners are Warned! John Lort Stokes and the H.M.S. Beagle in Australia 1837 - 1843*, Melbourne University Press, Melbourne 1989.

Horner, F., *The French Reconnaissance: Baudin in Australia 1801 - 1803*, Melbourne University Press, Melbourne, 1987.

Jukes, J.B., *Narrative of the Surveying Voyage of HMS Fly*, T. and W. Boone, London, 1847.

King, P.P., *Narrative of A Survey of the Intertropical and Western Coasts of Australia, performed between the years 1818 & 1822*, vols 1 & 2, John Murray, London, 1827. (Facsimile edn, Libraries Board of SA, 1969.)

Loney, J., and Stone, P., *Australia's Island Shipwrecks*, Neptune Press, Victoria, 1980.

Marchant, L.R.., *An Island Unto Itself: William Dampier and New Holland*, Hesperian Press, Perth, 1988.

Marchant, L.R., *France Australe*, Artlook Books, Perth, 1982.

Mulvaney, D.J., *Encounters in Place, Outsiders and Aboriginal Australians 1606-1985*, Queensland University Press, St. Lucia, 1989.

Reid, G., *From Dusk Till Dawn - A History Of Australian Lighthouses*, Macmillan, Melbourne, 1988.

Walsh, G.L., *Australia's Greatest Rock Art*. E.J. Brill/Robert Brown and Assoc. Bathurst, NSW, 1988.

Stokes, J.L., *Discoveries in Australia; with an account of the coasts and rivers explored and surveyed during the voyage of HMS Beagle, in the years [1837- 43]*. T.& W. Boone, London, 1846. (Facsimile edn, Libraries Board of SA, 1969.)

Houtman Abrolhos

——— *Abrolhos Islands Planning Strategy*. Final Report. Abrolhos Islands Consultative Committee and Abrolhos Islands Task Force, Geralton, WA, 1989.

——— *Batavia 1629—A Seventeenth Century Shipwreck*. Pamphlet. Western Australian Maritime Museum, Freemantle.

Fuller, P.J., Burbidge, A.A. and Owens, R., *Breeding Seabirds of the Houtman Abrolhos, Western Australia: 1991 - 1994*, Corella 18(4), 1994.

Hasluck, N., *The Bellarmine Jug*, Penguin, Melbourne, 1985.

Ingelman-Sundberg, C., *Relics from the Dutch East Indiaman, Zeewijk. Foundered in 1727*. Western Australian Museum Special Publication No.10, Perth, 1978.

Storr, G.M., Johnstone, R.E., and Griffin, P., *Birds of the Houtman Abrolhos, Western Australia*, Records of the Western Australian Museum, Supplement No.24, Perth, 1986.

Whyte, N. and Ralph, L., *Wrecks in the Houtman Abrolhos Islands*, Pamphlet, Western Australian Maritime Museum, Fremantle, 1993.

Bernier And Dorre Islands

Berry, P.F., Bradshaw, S.D. and Wilson, B.R., (eds.) *Research in Shark Bay, Report of the France-Australe Bicentenary Expedition Committee*, Western Australian Museum, Perth, 1990.

Jebb, M.A., 'The Lock Hospitals Experiment: Europeans, Aborigines and Venereal Disease' in Reece, B. and Stannage, T.,(eds), *European–Aboriginal Relationships in Western Australia*, Studies in Western Australian History VIII, University of Western Australia Press, Perth, 1984.

Short, J. and Turner, B., 'The Distribution and Abundance of the Banded and Rufous Hare-Wallabies, Lagostrophus fasciatus and Lagorchestes hirsutus,' *Biological Conservation* Vol. 60, 1992.

Short, J., 'Burrowing Bettong,' *Australian Natural History*, Vol. 24, No. 9, 1994.

Dampier Archipelago

Curry, S., 'William Dampier, Voyages to New Holland,' *Landscope* vol 8, No 4, 1993.

Virili, F.L., 'Aboriginal sites and rock art of the Dampier Archipelago, Western Australia,' in *Form in Indigenous Art: Schematisation in the Art of Aboriginal Australia and Prehistoric Europe*, ed. P.J. Ucko, Australian Institute of Aboriginal Studies, Canberra, 1977.

Christmas Island

——— *Christmas Island National Park, Plan of Management*, Australian Nature Conservation Agency, Canberra, 1994.

——— *Visitor Guide, Christmas Island National Park*, Indian Ocean, Australia. Pamphlet. Australian Nature Conservation Agency, Darwin.

Allen, G.R. and Steene, R.C., *Fishes of Christmas Island*, Christmas Island Natural History Association, Christmas Island, Indian Ocean, 1988.

Brett, D., 'Sea Birds in the Trees,' *Ecos* No. 61, CSIRO, Canberra, 1989.

Hicks, J., Rumpff, H. and Yorkston, H., *Christmas Crabs* (2nd ed.) Christmas Island Natural History Association, Christmas Island, Indian Ocean, 1990.

Reville, B.J., *A Visitor's Guide to the Birds of Christmas Island, Indian Ocean*. (2nd ed.) Christmas Island Natural History Association, Christmas Island, Indian Ocean, 1993.

Reville, B.J., Tranter, J.D. and Yorkston, H.D., *Conservation of the Endangered Seabird Abbott's Booby on Christmas Island.*

BIBLIOGRAPHY AND FURTHER READING

Reville, B.J., Tranter, J.D. and Yorkston, H.D., *Conservation of the Endangered Seabird Abbott's Booby on Christmas Island. The Monitoring Program 1983-89.* Occasional Paper No. 20. Australian National Parks and Wildlife Service, Canberra, 1990.

Bonaparte Archipelago - Bigge Island

Burbidge, A., McKenzie, N.L., and Kenneally, K.F., *Nature Conservation Reserves in the Kimberley,* Department of Conservation and Land Management, Perth, 1991.

Crawford, I.M., *The Art of the Wandjina,* Oxford University Press, Melbourne, 1968.

McGregor, A., and Chester, Q., *The Kimberley: Horizons of Stone,* Hodder & Stoughton, Sydney 1992.

Wessel Islands

Barrett, C., *Coast of Adventure, Untamed North Australia,* Robertson and Mullins, Melbourne, 1946.

Davis, S., *Above Capricorn, Aboriginal Biographies from Northern Australia,* Angus and Robertson, Sydney, 1994

Fisher, A. and Woinarski, J., 'Golden Bandicoot,' *Australian Natural History,* vol. 24, no. 11, 1994 - 95.

McIntosh, I., *The Whale and the Cross, Conversations with David Burrumarra MBE,* Historical Society of the Northern Territory, Darwin, 1994.

Woinarski, J. and Fischer, A. (eds.) 'The Wildlife of the Wessel Islands,' *Technical Report* No.60, Parks and Wildlife Commission of the Northern Territory, Darwin, 1996.

Sir Edward Pellew Group

——————— *Barranyi (North Island) National Park, Draft Plan Of Management,* Conservation Commission of the Northern Territory, 1994.

Bradley, J. J.(trans.), *Yanyuwa Country: the Yanyuwa people of Borroloola tell the history of their land,* Greenhouse Publications, Melbourne, 1988.

Johnson, K.A. and Kerle, J.A., *Flora And Vertebrate Fauna of the Sir Edward Pellew Group of Islands, Northern Territory,* Conservation Commission of the Northern Territory, Alice Springs, 1991.

Macknight, C.C., *The Voyage to 'Marege', Macassan Trepangers in Northern Australia,* Melbourne University Press, 1976.

Cape York and the Torres Strait

Mulrennan, M. and Hanssen, N., *Marine Strategy for the Torrres Strait, Policy Directions,* Australian National University North Australia Research Unit, Darwin, 1994.

MacGillivray, J., *Narrative of the Voyage of the HMS Rattlesnake, 1848 - 49 ,* T.W. Boone, London, 1852.

Prideaux, P., *From Spear to Pearl-shell. Somerset Cape York Peninsula 1864 - 1877,*

Boolarong Publications, Brisbane, 1988.

Sharp, N., *Footprints Along the Cape York Sandbeaches,* Aboriginal Studies Press, Canberra, 1992.

Walker, D., *Bridge And Barrier: The Natural and Cultural History of the Torres Strait,* Australian National University, Canberra, 1972.

Flinders Group

Cole, N., and David, B., 'Curious Drawings at Cape York Peninsula,' *Rock Art Research,* vol 9 no 1, 1992.

Hale, H. M. and Tindale, N. B., 'Aborigines of Princess Charlotte Bay, North Queensland,' *Records of the South Australian Museum 5,* 1933-1934.

Hinchinbrook Island

Thorsborne, A. and Thorsborne, M., *Hinchinbrook Island: The Land Time Forgot,* Weldon Publishing, Sydney, 1988.

Tracey, J.G., *The Vegetation of the Humid Tropical Region of North Queensland,* CSIRO, 1982.

Whitsunday Islands

Colefelt, D., *100 Magic Miles of the Great Barrier Reef,* Windward Publications, Sydney, 1993.

Fraser Island

Baverstock, F., *Fraser Island - Sands of Time,* ABC Enterprises, Sydney, 1985.

Sinclair, J., *Fraser Island and Cooloola,* Weldon Publishing, Sydney, 1990.

Capricorn Bunker Group

——————— *Exploring Capricornia, Great Barrier Reef Marine Park,* Great Barrier Reef Marine Park Authority/ Queensland National Parks and Wildlife Service, Rockhampton, 1993.

——————— *A Matter of Time, Sea Turtles in Queensland,* Queensland Dept. of Environment, Brisbane, 1994.

——————— *The Reader's Digest Book of the Great Barrier Reef,* Reader's Digest, Sydney, 1984.

Cribb, A.B. and Cribb, J.W., *The Plant Life of the Great Barrier Reef and Adjacent Shores,* University of Queensland Press, St Lucia, 1985.

Ward, W.T. and Saeger, P. (eds), *The Capricornia Section of the Great Barrier Reef, Past Present and Future,* Royal Society of Queensland and the Australian Coral Reef Society, Queensland Museum, Brisbane, 1984.

Wright, J., *The Coral Battleground,* Nelson, Melbourne, 1977.

Lord Howe Island

——————— *Lord Howe Island Permanent Park Preserve,* Plan of Management, Lord Howe Island Board and New South Wales National Parks and Wildlife Service, Sydney, 1986.

Hutton, I., *Discovering Australia's World Heritage: Lord Howe Island,* Conservation Press, Canberra, 1986

McDougall, I., Embleton, B.J. and Stone, D.B., 'Origin and Evolution of Lord Howe Island, Southwest Pacific Ocean' *Geological Society of Australia Journal,* 5, 1981.

Thompson, D., Bliss, P., and Priest, J., *The Geology of Lord Howe Island,* Map and Notes, Lord Howe Island Board and New South Wales Dept. of Mineral Resources, 1987.

Glennie Group - Great Glennie Island

Bowden, K.M., *George Bass 1771 - 1803* Oxford University Press, Melbourne, 1952.

Cumpston, J.S., *First Visitors to Bass Strait,* Roebuck, Canberra, 1973.

Jones, R., and Allen, J., 'A stratefied archeaological site on Great Glennie Island, Bass Strait', *Australian Archaeology* 9, 1979.

Kent Group - Deal Island

Brownrigg, M.B., *The Cruise of the 'Freak': A narrative of a visit to the islands in Bass and Banks Straits with some account of the islands,* Launceston, 1872.

Edgecombe, J., *Flinders Island and Eastern Bass Strait,* Author, Sydney, 1986.

Nixon, F.R., *The cruise of the 'Beacon' A narrative of a visit to the islands in Bass's Straits,* Bell & Daldy, London, 1857.

Turner, I., 'Artists' Camp Erith Island: March 1974,* *Overland 60,* Autumn 1975.

Hunter Group - Three Hummock Island

Alliston, E., *Escape to an Island,* Heinemann, Melbourne, 1966.

Alliston, E., *Island Affair,* Greenhouse, Melbourne, 1984.

Maatsuyker Group - Maatsuyker Island

——————— *Maatsuyker Island Conservation Area Management Strategies,* Tasmanian Parks and Wildlife Service, Australian Maritime Safety Authority, unpublished report, March 1993.

Gee, H., and Fenton, J. (eds), *The South West Book,* Australian Conservation Foundation, Melbourne, 1978.

Smith, S.J., and Banks, M.R., eds.) *Tasmanian Wilderness - World Heritage Values,* Royal Society of Tasmania, Hobart, 1993.

Stanley, K.M., *Guiding Lights,* St David's Park Publishing, Hobart, 1991.

White, G., *Islands of South-West Tasmania,* Self-published, Sydney, 1981

Macquarie Island

Bennett, I., *Shores Of Macquarie Island,* Rigby, Adelaide, 1971.

Copson, G. R. and Whinam, J. *Rabbits and Vegetation - Their Future on Macquarie Island,* Tasmanian Parks and Wildlife

Service, Hobart, 1994.

Cox, R., *"One of the Wonder Spots of the World..." Macquarie Island Nature Reserve,* Dept. of Parks Wildlife and Heritage, Hobart, 1990.

Evans, K., *Shipwrecks, Sealers and Scientists - A Guide to Sites of Cultural Heritage on Macquarie Island,* Tasmanian Parks and Wildlife Sevice, Hobart, 1995.

Grenfell Price, A., *The Winning of Australian Antarctica, Mawson's B.A.N.Z.A.R.E. Voyages 1929-1931,* Angus and Robertson, Sydney, 1962.

Mawson, D., *The Home of the Blizzard,* William Heinemann, London, 1915. (Abridged popular ed., Hodder and Stoughton, London, 1930)

Nuyts Archipelago - St Francis Island

Copley, P., *The Stick-nest Rats of Australia: Final Report to World Wildlife Fund (Australia),* Government Printer, Adelaide 1988.

Maguire, A., 'An account of a trip to St Francis Island in 1909,' *Proc. Royal Geo. Soc. Aust. SA Branch.* 21, 1920.

Archipelago of the Recherche - Mondrain Island

Horner, F., *Looking for La Perouse: D'Entrecasteaux in Australia and the South Pacific 1792 - 1793,* Melbourne University Press, Melbourne, 1995.

Serventy, V., 'Islands of Western Australia,' in Dutton, G. (ed), *The Book of Australian Islands,* Macmillan, Melbourne, 1986.

Heard Island

——————— *Heard Island Wilderness Reserve, Management Plan,* Australian Antarctic Division, Hobart, 1996.

Budd, G., 'Narrative, the Anare 1963 Expedition to Heard Island,' *ANARE Reports,* Series A, Vol. 1, Antarctic Division, Dept. of External Affairs, Melbourne, 1964.

Hughes, J.M.R., 'The Distribution And Composition Of Vascular Plant Communities On Heard Island,' *Polar Biology,* 7: 153-162, 1987

Scholes, A., *Fourteen Men: Story Of The Australian Antarctic Expedition To Heard Island,* Cheshire, Melbourne, 1949.

Thornton, M. (ed), *Heard Island Expedition 1983,* Spirit of Adventure, Sydney, 1983.

ACKNOWLEDGMENTS

This book would have remained a pipe-dream were it not for the extraordinary support of a great many people.

We are extremely appreciative of the generous assistance from Karyn Ashlin of Fuji Australia (Hanimex Industries) for film stock, Alan Ward of Vision Graphics for film processing, Christine Algie of C.R. Kennedy (agents for Asahi Pentax) and Ian Gibson of Paddy Pallin.

A number of individuals gave us valuable help and advice through the course of this project. These include: Martin Baird, Andrew Burbidge, Rob Easther, Pauline English, Tim Flannery, Robyn Graham, Robyn Grosvenor, Morea Grosvenor, Chris Henderson and Sally Hildred (whose yacht *Bydand* enabled us to reach several of the islands), Bert Hingley, Rob Jung, Bob Pascoe, Ian Roberts, Claire Robertson, Joc Schmiechen, Rob Tiley and James Woodford.

We also thank the following people and organisations who helped us during our visits to the islands:

Houtman Abrolhos: Andrew Burbidge and Phil Fuller of the Threatened Species and Communities Unit in CALM, Randall Owens of the WA Fisheries Department.

Bernier and Dorre Islands: Andrew Burbidge, Sue Hancock and Keith Morris of CALM, Jeff Short of CSIRO. *Bydand.*

Dampier Archipelago: Andrew Burbidge, Stefan Fritz and Peter Kendrick of CALM. *Bydand.*

Bigge Island: *Bydand.* Diane Deeming of Ansett Airlines.

Christmas Island: Graeme Beech, Roger Hart, Holger Rumpff and Max Orchard of ANCA, Richard Hill of RAOU, Margie Campbell.

Wessel Islands: James Barripang and family. Northern Land Council. John Woinarski and Aleric Fisher of the NT Parks and Wildlife Commission. Bob Pascoe. *Bydand.*

Sir Edward Pellew Group: Tim Hutton, Eddie Webber and Bryan Walsh of the NT Parks and Wildlife Commission. Alan Jupiter and the Baranyi Aboriginal Corporation, Bob Pascoe and Ian Roberts.

Cape York and the Torres Strait: Dennis Devine of the Queensland Department of Environment. *Bydand.*

Flinders Group: Dennis Devine of the Queensland Department of Environment. *Bydand.*

Hinchinbrook Island: Mark Burnham of the Queensland Department of Environment. Bill Pearce of Lucinda Wilderness Safaris.

Whitsunday Islands: Artie Jacobsen, Mark Luytens and Paul Harrison of the Queensland Department of Environment.

Fraser Island: Bay 4WD Hire.

Capricorn Bunker Group: John Olds, Jos Richter, Bruce Knuckey, John Meech and Greg Carter of the Queensland Department of Environment.

Lord Howe Island: Judy Mortlock and Dean Hiscox of the Lord Howe Island Board.

Cabbage Tree Island: Nicholas Carlile and David Priddel of the New South Wales National Parks and Wildlife Service.

Spectacle Island: Peter Short and Pauline Foote of the Dangar Island Bushfire Brigade.

Glennie Group: Jim Whelan of Wilsons Promontory National Park. Michael Glover of Department of Environment and Natural Resources, Victoria. Rob Jung.

Kent Group: Neil Smith of *Wild Wind.* The Hollier family.

Hunter Group: Neil Smith of *Wild Wind.* John and Eleanor Alliston.

Maatsuyker Group: Peter Mooney, Nigel Brothers, David Pemberton of the Tasmanian Parks and Wildlife Service. Owen Barret and David Clarke of the Australian Maritime Safety Authority. Robert Easther of *Domino.*

Macquarie Island: Geof Copson of the Tasmanian Parks and Wildlife Service. Jim Bleasel (former Director, Australian Antarctic Division). Robyn Graham.

Nuyts Archipelago: Peter Copley, David Armstrong, Robert Brandle, Peter Canty and Tony Robinson of the Resource Management Branch of the Department of Environment and Natural Resources, South Australia.

Archipelago of the Recherche: Bernie Haberley of CALM. *Wet Pussy.* Esperance Sea Rescue.

Heard Island: Josko Grubic of *Anaconda* and members of the 1983 Heard Island Expedition. Robyn Graham, Rod Ledingham, Atila Vrana and Rene Wanless of the Australian Antarctic Division.

Additional photographs were kindly made available by Simon Banks, Ian Brown, Clay Bryce, Andrew Burbidge, Peter Canty, Dave Carter, Greg Carter, Tony Fontes, Lincoln Hall, Rob Jung, Phillip Castleton, Jiri Lochman, Bob Mossel, Holger Rumpff, Ian Shaw, Steve Trémont, NT Parks and Wildlife Commission.

Helpful comments on aspects of the manuscript were provided by ANCA's Christmas Island staff, Andrew Burbidge, Mark Burnham, Nicholas Carlile, Geof Copson, Dennis Devine, Robyn Graham, Dean Hiscox, Artie Jacobsen, Rod Ledingham, Peter Ogilvie, John Olds, Randall Owens, Jos Richter, Tony Robinson, Eddie Webber, John Woinarski.

Finally, we are deeply grateful to Morea Grosvenor and Dale Arnott for their loyal support and encouragement throughout the project.

INDEX